Agile Certification Study Guide

Practice Questions for the PMI-ACP® exam and the Scrum Master Certification PSM I® exam

Part of the Agile Education Series™

By

Dan Tousignant, PMI-ACP, PSM-I, CSP, PMP

UPDATED JULY 2015

Published by:
Cape Project Management, Inc.
14 Bayview Ave
Plymouth, MA 02360
Visit us at http://www.CapeProjectManagement.com

Copyright © 2015 by Cape Project Management, Inc. and Daniel Tousignant

All rights reserved. No part of this book may be reproduced, stored in retrieval system, or transmitted in any form or by any means, electronic, mechanical, photocopying, recording, or otherwise, without the prior written permission from Cape Project Management, Inc.

"PMI®" and the PMI logo are service and trademarks registered in the United States and other nations.

"PMP®" and the PMP logo are certification marks registered in the United States and other nations.

PSM I is a trademark of Ken Schwaber and Scrum.org

"PMI-ACP" is a certification mark registered in the United States and other nations.

"The Agile Education Series" is a trademark of Cape Project Management, Inc.

All other brand or product names used in this guide are trade names or registered trademarks of their respective owners.

Printed in the United States of America

Second Printing, July 2015 – Version 2.1

ISBN 978-0-9848767-8-5

Table of Contents

Agile Certification Questions 4

Agile Certification Questions with Answers 91

Agile Glossary and Web Resources 188

Looking for online practice exams for Agile certifications? 241

References and Reccomended Reading 242

Agile Certification Questions

There are 371 practice exams in this study guide to help prepare you for the PMI-ACP exam. These questions are based upon the most recent PMI-ACP Exam content outline that went into effect on July 15, 2015. Questions 219-322 are specifically focused on Scrum and will also help you prepare for Scrum Master Certification: The PSM I assessment from Scrum.org

1. Which of the following is NOT a principle of the Agile Manifesto?

Choice
Working software is delivered frequently
Close, daily co-operation between business people and developers
Daily team meetings to review progress and address impediments
Projects built around motivated individuals, who should be trusted

2. A key component of Agile software development is:

Choice
Requirements should be complete before beginning development
Change must be minimized
Risk management is robust
Requirements are able to evolve during development

3. Which one is NOT a Pillar of Scrum?

Choice
Transparency
Inspection
Adaption
Empiricism

4. Which is NOT a Scrum Role?

Choice
Product Owner
Team Member
Project Manager
Scrum Master

Agile Certification Study Guide – Practice Questions

5. For a one month Sprint, what is the recommended duration of the Sprint Planning meeting?

Choice
2 sessions - 8 hours
1 Session - 4 hours
3 hours
Decided by the Team

6. What is the definition of velocity?

Choice
The number of Sprints per release
The number of items resolved in a daily Scrum
The number of story points completed in Sprint
The number of User Stories in a release

7. In which meeting do you capture lessons learned?

Choice
Sprint Planning
Sprint Review
Sprint Retrospective
Daily Status Meeting

8. Who developed extreme programming (XP)?

Choice
Mike Cohn
Ken Schwaber
Kent Beck
Alistair Cockburn

9. Which is NOT a role on an XP team?

Choice
Coach
Customer

Choice
Product Owner
Programmer

10. In Pair Programming, one programmer is responsible for all the coding in an iteration then the programmers switch for the next iteration.

Choice
True
False

11. Extreme Programming (XP) includes which of the following 3 practices?

Choice
Simple Design
Coding standards
Mange WIP
Sustainable Pace
Eliminate Waste

12. What is the role of the XP Coach?

Choice
Defines the right product to build
Determines the order to build
Ensures the product works
Helps the team stay on process

13. Which 2 are goals of Lean Software Development?

Choice
Ensure value
Simple Design
Continuous integration
Minimize waste

14. Which is NOT a principle of Lean?

Agile Certification Study Guide – Practice Questions

Choice
Eliminate waste
Time-box events
Deliver fast
Delay commitment

15. Kanban means _____ in Japanese?

Choice
User Story
Signal card
Visual card
Production Line

16. Which is the first step in setting up Kanban?

Choice
Place prioritized goals on the left column of the board
Decide on limits for items in queue and work in progress
Map your current workflow
Lay out a visual Kanban board

17. Which Agile framework adopts and tailors methods such as Scrum, Extreme Programming (XP), Agile Modeling (AM), Unified Process (UP), Kanban and Agile Data (AD) in order to support scaling.

Choice
DSDM
Crystal
Disciplined Agile Delivery (DAD)
Agile Delivery Framework (ADF)

18. Project X has an IRR of 12%, and Project Y has an IRR of 10%. Which project should be chosen as a better investment for the organization?

Choice
It depends on the payback period

Project Y
Project X
Project or Y, depending on the NPV

19. What is a product roadmap?

Choice
A list of reports and screens
A view of release candidates
Instructions for deployment
A backlog prioritization scheme

20. Your sponsor has asked for clarification on when releases of your product will ship and what those releases will contain. Which Agile deliverable would best answer their needs?

Choice
Product demo
Product roadmap
Product backlog
Sprint backlog

21. The acronym MoSCoW stands for a form of:

Choice
Estimation
Risk identification
Prioritization
Reporting

22. What is the reason to develop personas as part of User Story creation?

Choice
When the conversation is centered on the high-level flow of a process
When trying to better understand stakeholder demographics and general needs
When trying to capture the high-level objective of a specific

Agile Certification Study Guide – Practice Questions

requirement
When communicating what features will be included in the next release

23. When comparing communication styles, which of the following are true?

Choice
Paper-based communication has the lowest efficiency and the highest richness.
Face-to-face communication has the highest efficiency and the lowest richness.
Paper-based communication has the highest efficiency and the lowest richness.
Face-to-face communication has the highest efficiency and the highest richness.

24. Highly visible project displays are called:

Choice
Project radiators
Information refrigerators
Information radiators
Project distributors

25. Well-written User Stories that follow the INVEST model include which attributes?

Choice
Independent, Negotiable, Smart
Valuable, Easy-to-use, Timely
Negotiable, Estimable, Small
Independent, Valuable, Timely

26. Wireframe models help Agile teams:

Choice
Test designs
Confirm designs
Configure reports

Track velocity

27. Which of the following is the hierarchy of User Story creation?

Choice
Task, User Story, Feature, Theme
Theme, Epic, User Story, Task
User Story, Epic, Theme, Feature
Goal, Epic, Activity, User Story

28. Which of the following statements is true for measuring team velocity?

Choice
Velocity is not accurate when there are meetings that cut into development time.
Velocity measurements are disrupted when some project resources are part-time.
Velocity tracking does not allow for scope changes during the project.
Velocity measurements account for work done and disruptions on the project.

29. Self-organizing teams are characterized by their ability to:

Choice
Do their own filing
Sit where they like
Make local decisions
Make project-based decisions

30. High-performing teams feature which of the following sets of characteristics?

Choice
Consensus-driven, empowered, low trust
Self-organizing, plan-driven, empowered
Constructive disagreement, empowered, self-organizing
Consensus-driven, empowered, plan-driven

31. Which of the following sets of tools is least likely to be utilized by an Agile team?

Choice
Digital camera, task board
Wiki, planning poker cards
WBS, PERT charts
Smart board, card wall

32. Who typically has the best insight into task execution?

Choice
Project managers
Team members
Scrum Masters
Agile coaches

33. A servant leadership role includes:

Choice
Shielding team members from interruptions
Making commitments to stakeholders
Determining which features to include in an iteration
Assigning tasks to save time

34. All of the following are TRUE about communicating on distributed teams EXCEPT:

Choice
Should consider instant messaging tools
Should have an easier Storming phase
Need to spend more effort communicating
Have a higher need for videoconferencing

35. What is the sequence of Tuckman's stages of team formation and development progress?

Order
Storming
Performing
Norming

Forming

36. The Agile Manifesto states we value some items over others. Match the items in the columns below so each item on the left is valued over the corresponding item on the right.

	Choice
Individuals and Interactions	Following a Plan
Working Software	Comprehensive Documentation
Customer Collaboration	Processes and Tools
Responding to Change	Contract Negotiation

37. Your team is running three-week Sprints. How much time should you schedule for Sprint Review sessions?

Choice
1 hour, 15 minutes
45 minutes
3 hours
6 hours

38. A 4-hour Sprint Planning meeting is typical for a Sprint or Iteration that is how long?

Choice
Four weeks
Two weeks
One week
Four days

39. In Agile project management, responding to change is valued over _____ .

Choice
Contract negotiation
Following a plan
Customer collaboration
Processes and tools

40. The person responsible for the Scrum process, making sure it is used correctly and maximizing its benefit:

Agile Certification Study Guide – Practice Questions

Choice
Product Owner
Coach
Scrum Master
Scrum Manager

41. An artifact on an Agile project can best be described as:

Choice
A work output, typically a document, drawing, code, or model
The deliverable from a Sprint retrospective
The Agile model of persona
A document that describes how work the work needs to be done

42. Product Backlog Items (PBI) described as emergent, are expected to:

Choice
Become new technology requirements
Grow and change over time as users stories are completed
Become the highest priority items
Be the most recent stories added to the backlog

43. Which is a report of all the work that is "done?"

Choice
Burndown chart
Completion chart
Kanban chart
Sashimi

44. Traditional project management uses requirement decomposition. This can be comparable to _____ of Agile User Stories.

Choice
Sashimi
Definition of done
Disaggregation

De-separation

45. In Agile, _____ is the primary measure of progress:

Choice
Accelerated Burndown chart
Reduced risk
Increased customer satisfaction
Working software

46. User Stories are:

Choice
Negotiable
Baselined and not allowed to change
Created by the Agile Project Manager
The foundation of the roadmap

47. When maintaining the product backlog, this role represents the interests of the stakeholders, and ensures the value of the work completed:

Choice
Scrum Master
Agile Project Manager
Product Owner
Sponsor

48. Acronym describing the attributes of a product backlog:

Choice
DEEP
INVEST
ITERA
DESPO

49. Agile uses a number sequence for estimating. The series of numbers begin with 0, 1, 1 and are calculated by adding the previous two numbers to get the next number. This number sequence is called:

Choice
Sashimi
Velocity
Capacity
Fibonacci

50. This role would be responsible for determining when a release can occur:

Choice
Product Owner
Scrum Master
Development Team
Agile Project Manager

51. On an Agile team, the project leader works to remove impediments from blocking the team's progress. This is known as what type of leadership?

Choice
Servant
Command and control
Consensus-driven
Functional management

52. Which should NOT take place at the daily Scrum?

Choice
The Product Owner gives an update
The Scrum Master manages the time-box
The Development Team answers the three questions
Issues are raised and documented

53. What is the purpose of practicing asking the "5 Why's"?

Choice
To determine the scope of the Sprint
To determine the root cause of an issue
To determine the end result

Choice
To determine the prioritized backlog

54. Which of the following describe the roles in pair programming?

Choice
Pilot and the navigator
Driver and the navigator
Coder and the planner
Leader and the second chair

55. Which Agile methodology runs one week iterations; leverages the use of pair programming; and includes the roles of coach, customer, programmer, tracker, and tester?

Choice
Lean
Agile One
Scrum
XP

56. Continuous attention to _____ and good design enhances agility.

Choice
Best architectures
Technical excellence
Robust plans
Change control

57. The best description of a Sprint backlog is:

Choice
Daily progress for a Sprint over the Sprint's length
A prioritized list of tasks to be completed during the project
A prioritized list of requirements to be completed during the Sprint
A prioritized list of requirements to be completed for a release

58. This Agile methodology focuses on efficiency and habitability as components of project safety.

Choice
Scrum
Kanban
Extreme Programming
Crystal Clear

59. A time-boxed period to research a concept and/or create a simple prototype is called a(n):

Choice
Sprint
Iteration
Spike
Retrospective

60. In Scrum, who is responsible for managing the team?

Choice
Scrum Master
Project Manager
Development Team
Product Owner

61. Simplicity - the art of _____ - is essential

Choice
Maximizing the amount of work not done
Minimizing the amount of work done
Maximizing the customer collaboration
Minimizing contract negotiation

62. This approach includes a visual process management system and an approach to incremental, evolutionary process changes for organizations.

Choice
Kanban
Scrum

Extreme programming
Agile Unified Process

63. Which following statement is the least accurate regarding the Burndown chart?

Choice
It is calculated using hours or points
It is updated by the development team daily
It provides insight into the quality of the product
It reflects work remaining

64. What can be described as "one or two written sentences; a series of conversations about the desired functionality."

Choice
User Story
Story point
Epic
Product roadmap

65. This artifact contains release names with expected dates and includes major features, client-side impacts, server-side applications, platform support and markets served:

Choice
Risk Burndown graph
Product roadmap
Release migration plan
Product vision

66. Which statement is least accurate when providing a definition of "Done"?

Choice
It is the exit criteria to determine whether a product backlog item is complete
It may vary depending on the project
It is defined by the Scrum Master
It becomes more complete over time

67. All of the following are among the seven principles of the Lean approach with the exception of:

Choice
Amplified learning
Decide as late as possible
Build integrity in
Maximize the work performed

68. Typically calculated in story points, this is the rate at which the team converts "Done" items in a single Sprint:

Choice
Burndown rate
Burn-up rate
Velocity
Capacity

69. All of the following occur in the second half of the Sprint planning meeting EXECPT:

Choice
The Development Team identifies improvements that it will implement in the next Sprint
The Product Owner is answers questions and clarifies User Stories
The Development Team commits to work in the Sprint
Tasks are defined for the User Stories

70. The best architectures, requirements, and designs emerge from:

Choice
Hand-picked teams
Co-located teams
Self-organizing teams
Cross-functional teams

71. The Sprint retrospective:

Choice
Is intended to promote continuous process improvements

Choice
Is held at the end of each release
Is conducted to provide the sponsor with key information on team progress
Is optional

72. An iteration prior to a release that includes final documentation, integration testing, training and some small tweaks is called:

Choice
Hardening Iteration
Buffer Iteration
Release Iteration
Integration Iteration

73. Analyzing the current organizational processes, per project requirements, and making needed process changes is called:

Choice
Value Stream Mapping
Release Planning
Use Case Development
Process Tailoring

74. At the end of first iteration, the team finishes User Stories A, B and 50% of C. What is the team velocity?

> The story sizes were:
> Story A = 8 Points
> Story B = 1 Points
> Story C = 5 Points
> Story D = 3 Points

Choice
11.5
9
14
16

75. Suppose 8 new members joined the development team, and the team size is now 15. The daily Scrum is getting noisy and exceeding the 15 minutes time-box. What is the most effective way to address this situation?

Choice
Divide the team into two teams with minimum dependency and have two separate daily Scrums.
Do nothing; allow the large team to exceed the time-box by a few minutes each meeting.
Increase the time-box for the daily Scrum to 30 minutes.
Ask the team members to only update on the impediments and highlight only the important ones.

76. While developing a story during the iteration, team discovered new tasks that were not identified earlier. A newly discovered task is such that the User Story cannot be completed during the iterations. What is the most appropriate action for the team to perform?

Choice
Let the Product Owner decide if there is still a way to meet the iteration goals.
Discuss the situation with the Scrum Master and see if there is still a way to meet the iteration goals.
Drop the User Story and inform the Product Owner that it will be delivered in the next iteration.
Modify the scope of other User Stories to allow completion of the Sprint backlog.

77. Suppose your team velocity is 8 story points, and the product backlog items are ordered by priority as shown below. If you are in a Sprint Planning meeting and need to commit to the User Stories for the next iteration, which ones will you select?

   Story 1 = 3 Story Points
   Story 2 = 4 Story Points
   Story 3 = 3 Story Points
   Story 4 = 1 Story Points

Choice
Story 1, 2, and 3
Story 1, 2, and 4
Story 1 and 2

Story 2, 3, and 4

78. During Sprint Planning, the architect is constantly estimating higher than everyone else, and team members are increasing their estimates to accommodate her. This is an example of:

Choice
Dominating
Influencing
Dictating
Anchoring

79. Traditional Project Management uses the Work Breakdown Structure to develop requirements in terms of activities. What type of breakdown structure is used in Agile for this purpose?

Choice
Feature Breakdown Structure
Product Roadmap
Product Breakdown Structure
Sprint Backlog

80. The Agile artifact that describes the incremental nature of how a product will be built and delivered over time, along with the important factors that drive each individual release is called:

Choice
Product Vision Statement
Product Wireframe
Product Roadmap
Product Theme

81. Suppose you are performing integrated testing on each of the different product increments developed during an iteration to ensure that the increments work together as a whole. What type of iteration is this?

Choice
Hybrid Iteration
Hardening Iteration

Choice
Heuristic Iteration
Handoff Iteration

82. Which of the following is an Agile practice promoted by XP that is often used in conjunction with other Agile methods?

Choice
Dynamic Systems Development Method (DSDM)
Feature Driven Development (FDD)
Adaptive Software Development (ASD)
Test Driven Development (TDD)

83. All of the following are metrics used in Agile except:

Choice
Real Time
Velocity
Escaped defects
Cycle time

84. Alistair Cockburn created the Crystal Family of Agile Methods, all of whose names contain colors of quartz crystals taken from geology. What two characteristics of an Agile project are used to determine the color of the Crystal method?

Choice
Size and complexity
Duration and complexity
Size and criticality
Duration and criticality

85. When implementing Agile Project Management, risk management that occurs simply by following Agile best practices such as iterative planning and review activities is called:

Choice
Inherent risk management
Organic risk management
Overt risk management

Intrinsic risk management

86. One of the major tools and techniques used in Lean Software Development is Value Stream Mapping. What is the primary purpose of value stream mapping?

Choice
To improve business processes
To identify and eliminate waste
To ensure product quality
To increase customer value

87. Which of the following is NOT recognized as a "unit" that can be used for estimating the size of the requirements on your Agile project?

Choice
Real time
Relative size
Ideal time
Ideal size

88. Extreme Programming (XP) defines four basic activities that are performed during the software development process. These include designing, coding, testing and ... ?

Choice
Collaborating
Leveling
Communicating
Listening

89. Agile project management and product development use several types of documents specific to each iteration; they are known as "artifacts." All of the following documents are Agile iteration artifacts except:

Choice
Iteration Vision Statement
Iteration Backlog
Iteration Plan
Iteration Burndown Chart

Agile Certification Study Guide – Practice Questions

90. In Sprint Planning, the development team estimates User Stories provided by the Product Owner and agrees to ones that can be completed in the Sprint. This is an example of:

Choice
Osmotic Communication
Reciprocal Commitment
Universal Acceptance
Tacit Agreement

91. Which of these statements is NOT correct about Ideal time and Calendar time?

Choice
Ideal time is the time that is actually required to complete the work
Both of them convey the same meaning
Calendar time is the amount of time that passes on clock (calendar days)
Normally calendar days are not equal to ideal days

92. Which of the following is NOT an Extreme Programming Practice?

Choice
Pair Programming
Planning Game
Test Driven Development
Extreme Persona

93. On XP teams, what is expected from the Project Manager?

Choice
Coach the team on Agile practices
Help the team work with the rest of the organization
Provide domain expertise to the team
Define the software

94. Technical debt is the total amount of less-than-perfect _____ in your project.

Choice
Defects

Design and implementation decisions
Code commenting
Code Sharing

95. _____ an estimate refers to estimating a story based on its relationship to one or more other stories.

Choice
Triangulating
Triaging
Aggregating
Disaggregating

96. Which type of estimation refers to splitting a story into smaller and easier to manage pieces?

Choice
Expert opinion
Analogy
Disaggregation
Bottom up

97. The _____ the technical debt means the _____ the intrinsic quality?

Choice
higher, higher
higher, lower
lower, lower
lower, higher

98. A technique in which a team collaboratively discusses acceptance criteria and then distills them into a set of concrete tests before development begins is called:

Choice
Feature Driven Development (FDD)
Acceptance Test Driven Development (ATDD)
Test Driven Development (TDD)
User Story workshops

Agile Certification Study Guide – Practice Questions

99. Setting up development work in a way that the team can figure out what to do next is called:

Choice
A pull system
Push system
Critical path
Sprint backlog

100. What Agile development approach is being used when the whole team works towards solving a complex problem?

Choice
Swarming
Norming
Performing
Collaborating

101. The acronym for a good product backlog is DEEP. What does it stand for?

Choice
Detailed appropriately, Emergent, Estimated, and Practical
Detailed appropriately, Emergent, Estimated, and Prioritized
Descriptive, Emergent, Estimated, and Prioritized
Detailed appropriately, Exact, Estimated, and Prioritized

102. Which of the following is NOT a characteristic of an Agile plan?

Choice
Follows rolling wave planning approach
Are top down
Easy to change
Shows dependencies of one task to others

103. Which type of risk analysis does an Agile team use to identify risks on their project?

Choice
Risk Burndown Chart

Choice
Pareto Analysis
Qualitative Risk Analysis
Quantitative Analysis

104. The trend of work remaining across time in a Sprint, a release, or a product, with work remaining tracked on the vertical axis and the time periods tracked on the horizontal axis is called a _____.

Choice
Burndown Chart
Burnup Chart
Progress Chart
Parking Lot Chart

105. Which list below includes the attributes of a good User Story?

Choice
Small, estimable, independent, negotiable
Testable, estimable, renewable, valuable
Negotiable, small, explainable, valid
Valuable, estimable, dependent, small

106. The Agile Manifesto principle, "Our highest priority is to satisfy the customer through early and continuous delivery of valuable software," is achieved through which Scrum practice?

Choice
Daily Scrum
Sprints
Release Planning
Sprint Planning

107. During which meeting do team members synchronize their work and progress and report any impediments to the Scrum Master for removal?

Choice
Sprint Planning meeting
Daily Scrum

Sprint Retrospective
Weekly Status meeting

108. Who is responsible for change management in Scrum projects?

Choice
Project Manager
Project Sponsor
Scrum Master
Product Owner

109. The technique used to analyze the flow of information and materials through a system to eliminate waste is:

Choice
Fishbone diagramming
Flow charting
Value stream mapping
Pareto analysis

110. Which of the following defines the goal of testing in Lean software development?

Choice
Testing is to improve the process and quality
Testing plays the most crucial role in ensuring the intrinsic value of the product
In Lean software development, testing always refers to Test Driven Development
Testing is primarily performed to ensure that we don't have escaped defects

111. In Kanban, a diagram that describes the overall flow and provides a measurement for every significant step in the workflow is the:

Choice
Kanban board
Cumulative Flow diagram
Burn Down Chart

Choice
Parking Lot Chart

112. A visual control used in Lean software development to represent the velocity of business solutions delivered over time is a:

Choice
Business Value Delivered Chart
Burndown Chart
Cumulative Flow Diagram
Burn-Up chart

113. Suppose the current release will be complete after 6 two-week iterations. A team member is insisting that the Product Owner write the acceptance test cases for the entire release. How should the Scrum Master respond?

Choice
Since the release duration is not very long, agree with the idea and explain the advantage of doing this to Product Owner.
It's up to the team and Product Owner to determine if it is a good idea.
Explain that stories should only be discussed in detail prior to development. Elaborating stories not in the next couple of iterations is wasteful.
Decide to dedicate first Sprint to elaborating all the stories

114. Suppose your team is working to create a commercial website and is in the process of developing User Stories by role. One team member suggests rather than only thinking about the target user, we should also think of some exceptional users who use the system very differently. What type of user is this team member referring to?

Choice
High Risk User
Admin Role
Extreme Persona
Performance Tester

115. Which of the following testing is done manually by the XP team?

Choice
Regression testing

Unit testing
Integration testing
Exploratory testing

116. Which of the following are 2 attributes of Exploratory testing?

Choice
It involves minimum planning and maximum test execution
It is typically automated
It is unscripted testing
It is often the sole testing technique

117. A change made to the internal structure of software that makes it easier to understand and cheaper to modify but does not change its observable behavior is referred to as:

Choice
A Spike
Technical debt
A User Story
Refactoring

118. An iteration takes place in a time frame with specific start and end dates, called a time-box. Which of the following is NOT an advantage of time-boxing?

Choice
Establishes a WIP limit
Forces prioritization
Demonstrates progress
Helps control technical debt

119. Match each component of the Agile Triangle (on the left) to its associated description (on the right).

	Choice
Value	Cost, Schedule, Scope
Quality	Releasable Product
Constraints	Reliable, Adaptable Product

Agile Certification Study Guide – Practice Questions

120. Identify the three components of the Agile Triangle.

Choice
Quality
Value
Cost
Constraints
Scope
Leadership

121. There are four critical actions that should be embraced by an adaptive leader: improving speed-to-value, having a passion for quality, doing less, and _____.

Choice
Inspiring staff
Managing conflict
Facilitating meetings
Ensuring effective communication

122. The 3 items are required for an Agile, adaptive environment:

Choice
People
Process
Product
Tools
Technology

123. What is the name of this facilitated process? One or more team members sequence the product backlog from smallest to largest User Story. The rest of the team validates the sequence. The whole team uses a sizing method such as T-shirt size or Fibonacci sequence to group the user stories.

Choice
Relative estimation
Pairwise comparison
Planning Poker

Affinity estimating	

124. While managing the Agile Product Lifecycle, Match the frequency with which you update the five Agile plans.

	Choice
Product Roadmap	Daily by the individual
Sprint Plan	Semi-Annually by The Product Owner
Daily Plan (Scrum)	Annually by Product Owner
Release Plan	Each iteration by the team
Product Vision	Quarterly by the Product Owner and teams

125. What are the 5 values of Agile Modeling?

Choice
Communication, Simplicity, Feedback, Courage, Humility
Communication, Efficiency, Transparency, Courage, Humility
Communication, Simplicity, Feedback, Adaptation, Continuous Improvement

126. Pick the one factor that is NOT part of the Agile Scaling Model.

Choice
Team size
Geographical distribution
Domain complexity
Leadership Style
Organizational distribution
Technical complexity
Organizational complexity

127. What are the 5 phases of the Agile Project Management (APM) delivery framework?

Choice
Envision, Speculate, Explore, Adapt, Close
Elaborate, Speculate, Examine, Adapt, Close

Envision, Enable, Explore, Enhance, Close

128. The ultimate goal of _____ is to deploy all but the last few hours of work at any time.

Choice
Continuous Integration
Collective Code Ownership
Synchronous Builds
Asynchronous Builds

129. When reading a burn-down chart, what does each status measurement say about project performance? Match the items below.

	Choice
Actual Work Line is above the Ideal Work Line	Ahead of Schedule
Actual Work Line is below the Ideal Work Line	On Schedule
Actual Work Line is on the Ideal Work Line	Behind Schedule

130. How do you read a burndown bar chart? Match the phrases below to create instructions.

	Choice
As tasks are completed	the bottom of the bar is raised.
When tasks are added to the original set	the top of the bar is lowered.
When tasks are removed from the original set	the top of the bar moves up or down.
When the amount of work involved in a task changes	the bottom of the bar is lowered.

131. What are the three components of an Agile Project Charter?

Choice
Success Criteria
Vision
Objectives

Scope
Mission
Risks

132. Match each activity (on the left) to its definition (on the right).

	Choice
Communication	The developer submits code for testing. The UX designer checks that the developer implemented the elements correctly.
Coordination	Pair programming
Cooperation	A slide presentation by the Product Owner to stakeholders
Collaboration	The Product Owner adjusts some story priority to meet the dependency of another team.

133. In XP, the practice that any developer can change any line of code to add functionality, fix bugs, improve designs, or refactor demonstrates:

Choice
Collective Code Ownership
Source Code Control
Pair Programming
Continuous Integration

134. When the Agile team works in a single location, the team is said to be _____.

Choice
Co-located
Distributed
Outsourced
Functional

135. Teams of members working in different physical locations are called:

Choice
Co-located Teams
Distributed Teams

Outsourced Teams
Global Teams

136. On Agile teams, conflict is to be avoided at all cost.

Choice
True
False

137. Suppose you are a Scrum Master on a new Agile team. Which of the following strategies is best way to resolve conflict on the team?

Choice
Collaborate
Negotiate
Smooth over
Ignore
Use your authority

138. Match the response options to each level of conflict.

	Choice
Level 1: Problem to Solve	Establish safe structures again
Level 2: Disagreement	Accommodate, negotiate, get factual
Level 3: Contest	Do whatever is necessary
Level 4 : Crusade	Support and safety
Level 5 : World War	Collaboration or consensus

139. Which of the following is NOT one of the 5 common conflict types?

Choice
Compensation anxiety
Lack of role clarity
Difference in prioritizing tasks
Working in silos
Waiting on completion of task dependencies
Lack of communication

140. When we use the term "container" in Scrum what are we referring to?

Choice
A Sprint or Iteration
Source code repository
Development team room
A vertical slice of functionality

141. Prioritize from high to low the sequence of User Story development.

Order
Lower-value, low-risk
High-value, low-risk
High-value, high-risk stories
Low-value, high-risk

142. The number of days needed between feature specification and production delivery is called:

Choice
Cycle time
Real time
Ideal time
Calendar time

143. The PM Declaration of Interdependence is a set of six management principles initially intended for project managers of Agile software development projects. Match the items below to identify the principles.

	Choice
We deliver reliable results	through group accountability for results and shared responsibility for team effectiveness
We expect uncertainty	through situationally specific strategies, processes and practices
We boost performance	by making continuous flow of value our focus
We improve effectiveness and reliability	by engaging customers in frequent interactions and shared ownership

We unleash creativity and innovation	and manage for it through iterations, anticipation, and adaptation
We increase return on investment	by recognizing that individuals are the ultimate source of value.

144. All of the following are attributes of the definition of "Done", EXECPT:

Choice
It is a static artifact
It is an audible checklist
It is a primary reporting mechanism for team members on User Story progress
It is crucial to a high-performing team

145. The way that we calculate the number of years it takes to break even from undertaking a project which also takes into account the time value of money is the:

Choice
Pay-back period
Discounted pay-back period
NPV
Cumulative cash flow

146. DRY is an acronym for which Agile development principle?

Choice
Development Requires You
Don't Repeat Yourself
Deploy RepeatedlY
Develop, Refactor, Yagni

147. When is the best time to perform Earned Value Measurement (EVM) in Agile projects?

Choice
After the iteration
After a release
During an iteration

Agile Certification Study Guide – Practice Questions

Choice
Never - we don't perform EVM in Agile

148. Emotional intelligence includes all of the following except:

Choice
Self-awareness
Motivation
Commitment
Influence
Intuitiveness
Conscientiousness

149. In a burndown chart, if the remaining work line is above the expected work line, what does this signify?

Choice
The project is ahead of schedule
The project is behind schedule
The resources are performing above expectation
The project is being well managed

150. Empirical process control constitutes a continuous cycle of inspecting the process for correct operation and results and adapting the process as needed. What characteristics does this apply to in Scrum?

Choice
Self-organization, Collaboration and Time-boxing
Quality, Cost and Scope
Scrums, Sprint and Releases
Transparency, Inspection and Adaptation

151. Bugs reported by the customer that have slipped by all software quality processes are represented in this metric.

Choice
Technical debt
Escaped defects
Risk burndown

Code quality

152. Testing that often occurs between "Done" and "Done, Done" is:

Choice
Exploratory testing
Acceptance testing
Unit testing
Test driven development

153. Which of the following is NOT one of the 12 core practices of XP:

Choice
The Planning Game
Planning Poker
Small Releases
System Metaphor

154. Which of the following is NOT one of the 12 core practices of XP:

Choice
Simple Design
Continuous Testing
Vertical Slicing
Refactoring

155. Which of the following is NOT one of the 12 core practices of XP:

Choice
Pair Programming
Collective Code Ownership
40-Hour Work Week
Minimize Waste

156. Which of the following is NOT one of the 12 core practices of XP?

Choice
Visualize the flow

On-site Customer
Coding Standards
System Metaphor

157. Which of the following is NOT a reason to use a Feature Breakdown Structure (FBS) instead of a Work Breakdown Structure (WBS)?

Choice
It allows communication between the customer and the development team in terms both can understand.
It allows you to baseline your project plan due to absence of change.
It allows tracking of work against the actual business value produced.
It allows the customer to prioritize the team's work based on business value.

158. Team members who are part-time on your project will see at least a 15% reduction in their productivity per hour. The type of resource model in Agile is called:

Choice
Collocated
Fractional assignments
Distributed resources
Over-allocated resources

159. Frequent verification and validation is key in Agile but each approach produces a very different result. Verification determines _____ whereas validation determines _____.

Choice
if the product is "done" | if the product is "done - done"
if I am I building the product right | if I am I building the right product
if I am I building the right product | if I am I building the product right
if the product has passed unit testing | if the product has passed acceptance testing

160. What type of time estimation excludes non-programming time?

Choice
Ideal Time
Calendar time
Duration
Real Time

161. Which of the following is an example of an information radiator?

Choice
An email of a status report
A text of a quick question to the Product Owner
A whiteboard showing the state of work
A face-to-face conversation

162. Assuming all projects require the same amount of up-front investment, the project with the highest _____ would be considered the best and undertaken first.

Choice
Earned Value Management (EVM)
Internal Rate of Return (IRR)
Net Present Value (NPV)
Budget at Completion (BAC)

163. Match each Agile requirement type (on the left) to its definition (on the right).

	Choice
Feature	Fundamental unit of work that must be completed to make a progress on a Story
User Story	Any requirement that is NOT a User Story (e.g. technical enabling, analysis, reminder to have conversation)
Story	Business solution, capability or enhancement that ultimately provides value to the business.
Task	Describes the interaction of the users with the system

164. 80% of the value comes from 20% of the work. Which law is this referring to?

Choice
Parkinson's Law

Moore's Law
Pareto's Law
Jevon's Paradox

165. Based on the following information, determine the number weeks until the next release.

    Length of a Sprint = 2 weeks
    Velocity of team = 35 points
    Number of story points assigned to minimum marketable features (MMF) = 280 points

Choice
8 weeks
12 weeks
16 weeks
9 weeks

166. What is the correct sequence of activities in release planning?

Order
Create release plan by date or scope
Estimate the cost of the features
Split features using the MMF perspective
Estimate the value of the features
Prioritize features
Write stories for features
Identify features

167. Sequence the following concepts to create the popular acronym for creating good User Stories.

Order
Valuable
Small
Negotiable
Independent
Testable

Estimable	

168. Match the definitions (on the right) to each of the characteristics of a good User Story (on the left).

	Choice
Independent	If a story does not have discernable value it should not be done
Negotiable	Each story needs to be proven that it is "done"
Valuable	User Stories average 3-4 days of work
Estimable	Stories can be worked on in any order
Small	A story is not a contract
Testable	A story has to be able to be sized so it can be properly prioritized

169. At minimum, all Kanban boards should have the following columns:

Choice
To-Do, Doing, Done
Analysis, Design, Develop, Test, Deploy
Backlog, Design, Develop, Unit Test, Acceptance Test, Ready-to-ship
The Kanban columns are determined by the team

170. The Kano Model supports what Agile planning activity?

Choice
Estimation
Prioritization
Sizing
Continuous Integration

171. Which one is NOT a level of need in the Kano Model?

Choice
Basic Needs
Performance Needs
Enabling Needs
Excitement Needs

172. Match a definition (on the right) to a Lean principle (on the left).

	Choice
Eliminate Waste	Refactor - eliminate code duplication to zero
Create Knowledge	Limit work to capacity
Build Quality In	Maintain a culture of constant improvement
Defer Commitment	Create nothing but value
Optimize the Whole	Focus on the entire value stream
Deliver Fast	Schedule irreversible decisions at the last responsible moment

173. Net present value (NPV) is a ratio that compares the value of a dollar today to the value of that same dollar in the future. An NPV that is negative suggests what?

Choice
The project should be rejected
I don't have enough information
The project should be deferred
The project should be put on hold until the value is 0

174. What is the Agile Open Space concept?

Choice
When cubicles walls are removed for an Agile team.
It is a meeting designed to allow Agile practitioners to meet in self-organizing groups where they can share their latest ideas and challenges.
The choice to collocate all team members for the beginning of a project.
It is a core principle of ADSM

175. Osmotic communication is when team members obtain information from overheard conversations.

Choice
True
False

176. The 80/20 rule is also known as what law?

Choice
Little Law
Pareto's Law
Mohr's Law
The Law of Averages

177. Pareto Analysis is the exercise of determining what 40% of functionality can be delivered with 60% of the effort.

Choice
True
False

178. The length of time to recover the cost of a project investment is the:

Choice
Net Present Value
Payback Period
Earned Value
ROI

179. An archetypal user of a systems is called a(n):

Choice
Super user
Admin
Persona
UX engineer

180. Personas are used in Agile requirements to depict which type of user?

Choice
Real users
Fictitious users
Super users
Beta testers

181. When is Planning Poker used?

Choice
During backlog prioritization
As part of Pareto Analysis
During User Story sizing and estimating
As part of the Daily Stand-up

182. Process Tailoring is the iterative approach implementing your SDLC process.

Choice
True
False

183. Product roadmaps are more accurate the closer we get to an actual release.

Choice
True
False

184. Which is the process of continuously improving and detailing a plan as more detailed and specific information and more accurate estimates become available as the project progresses?

Choice
Process Tailoring
Pareto Analysis
Progressive Elaboration
Open Space Planning

185. What is a change made to the internal structure of software to make it easier to understand and cheaper to modify without changing its observable behavior?

Choice
Pair Programming
Continuous Improvement
Test Driven Development
Refactoring

186. Refactoring is a key way of preventing technical debt.

Choice
True
False

187. Which Agile estimation technique is based upon relative sizing?

Choice
Ideal time
Bottom up
Story points
Little's Law

188. What kind of User Story is written to provide an opportunity to research a solution in order to provide an estimate?

Choice
Sprint Story
Persona Story
Spike Story
Needle Story

189. Sequence the activities that occur in a Retrospective meeting.

Order
Generate Insights
Set the Stage
Close the Retrospective
Decide What to Do
Gather Data

190. Triple Nickels is a technique used in what kind of meeting?

Choice
Sprint Planning
Daily Scrum
Sprint Retrospective
XP Planning Game

Agile Certification Study Guide – Practice Questions

191. Who is responsible for managing ROI in Agile projects?

Choice
The Project Sponsor
The Product Owner
The Agile Project Manager
The Scrum of Scrums Master

192. The purpose of the Scrum of Scrums is to perform what function?

Choice
To increase knowledge of Agile within the organization
To provide dashboard reporting to executives
To manage cross-team dependencies working on the same project or product
To ensure team building and staff development occurs

193. Which one is NOT one of the 5 common risk areas mitigated by Agile?

Choice
Intrinsic schedule flaw
Specification breakdown
Scope creep
Stakeholder apathy
Personnel loss
Productivity variation

194. In what order do your select requirements to work on in a risk adjusted backlog?

Order
Low Risk High Value
High Risk Low Value
High Risk High Value
Low Risk Low Value

195. Which Agile method promotes the practice of risk-based Spike or Spike solutions?

Choice
Scrum
AgileUP
ADSM
Extreme Programming

196. Which one is NOT a reason to perform a Spike?

Choice
To perform basic research to familiarize the team with a new technology or domain
To analyze the expected behavior of a large story so the team can split the story into estimable pieces.
To defer a story until a later Sprint while still showing progress to the Product Owner
To do some prototyping to gain confidence in a technological approach

197. Which artifact is useful for seeing total project risk increasing or decreasing over time?

Choice
Burndown bar chart
Risk Burn-Up chart
Risk Burndown Graph
Risk Map

198. On a risk map or a risk heat map, the vertical and horizontal axes represent:

Choice
Effort and Impact
Probability and Impact
Probability and Exposure
Impact and Exposure

199. The Project Leader's primary responsibilities are to "move boulders and carry water." What is this an example of?

Choice
Servant leadership
Leadership by example
Command and control leadership
The leadership metaphor

200. In XP, what is the practice of creating a story about a future system that everyone - customers, programmers, and managers - can tell about how the system works?

Choice
Extreme persona
Wireframe
System metaphor
Simple design

201. What Agile requirements management approach displays a roadmap using the following approach?

    The horizontal axis shows a high level overview of the system under development and the value it adds to the users.

    The vertical axis organizes detailed stories into releases according to importance, priority, etc.

Choice
Release Planning Matrix
User Story Map
Agile Requirements Map
User Story Burndown Map

202. Which XP practice promotes the restriction on overtime?

Choice
Sustainable Pace
Pair Programming
Servant Leadership
Small Releases

203. What is the Agile term for the time period when some or all of the following occur: beta testing, regression testing, product integration, integration testing, documentation, defect fixing.

Choice
Spike
Code Freeze
Tail
Lag

204. Agile development prevents technical debt.

Choice
True
False

205. In Agile development, what is the term for the internal things that you choose not to do now, knowing they will impede future development if left undone?

Choice
Escaped defects
Verification and validation results
Technical debt
Intrinsic quality

206. What is the purpose of running a test before you develop the code?

Choice
To complete all test cases
To ensure it fails
To ensure it passes
To be cross-functional

207. Match the time box to the Scrum meeting for a one-month Sprint.

	Choice
Daily Scrum	3 hours
Sprint Review	15 Minutes
Sprint Planning	4 hours

Sprint Retrospective	8 Hours

208. A reminder for the developer and Product Owner to have a conversation is:

Choice
The Sprint planning meeting
Backlog grooming
A User Story
An Agile reminder

209. Wideband Delphi is used by an Agile Project manager to support what activity?

Choice
Prioritization
Scheduling
Estimation
Risk Management

210. The purpose of Work in Progress (WIP) limits is to prevent the unintentional accumulation of work, so there isn't a bottleneck.

Choice
True
False

211. Which 5 roles are defined by Extreme Programming?

Choice
Scrum Master
Coach
Customer
Stakeholder
Programmer
Tracker
Product Owner
Tester

Agile Certification Study Guide – Practice Questions

212. Simple Design, Pair Programming, Test-Driven Development, Design Improvement are all practices of which Agile methodology?

Choice
Scrum
Feature Driven Development (FDD)
Extreme Programming (XP)
Dynamic Systems Development Method (DSDM)
Crystal Clear
Rational Unified Process (RUP)
Agile Unified Process (AgileUP)

213. Which of the following is not an Agile methodology?

Choice
Scrum
Feature Driven Development (FDD)
Extreme Programming (XP)
Dynamic Systems Development Method (DSDM)
Program Evaluation Review Technique (PERT)
Crystal Clear
Rational Unified Process (RUP)
Agile Unified Process (AgileUP)

214. Incremental delivery means:

Choice
Deliver nonfunctional increments in the iteration retrospectives.
Release working software only after testing each increment.
Improve and elaborate our Agile process with each increment delivered.
Deploy functional increments over the course of the project.

215. When we practice active listening, what are the levels through which our listening skills progress?

Choice
Global listening, Focused listening, Intuitive listening
Interested listening, Focused listening, Global listening
Self-centered listening, Focused listening, Intuitive listening
Internal listening, Focused listening, Global listening

216. Which Agile method goes through the following stages:

    1a. The Feasibility Study
    1b. The Business Study
    2. Functional Model Iteration
    3. System Design and Build Iteration
    4. Implementation

Choice
Rational Unified Process (RUP)
Feature Driven Development (FDD)
Dynamic Systems Development Method (DSDM)
Lean Software Development
Scrum

217. What is a Japanese term used in Lean software development is an activity that is wasteful, unproductive, and doesn't add value?

Choice
Sashimi
Kanban
Muda
Kairoshi

218. Which one is not a value of Lean Development?

Choice
Pursue perfection
Ensure collective code ownership
After a project flows, keep improving it
Balance long-term improvement and short-term improvement

219. Pick 5 activities that are the responsibilities of the development team in Scrum.

Choice
Provides estimates
Prioritizes the backlog
Creates User Stories
Commits to the Sprint
Performs user acceptance
Facilitates meetings
Champions Scrum
Volunteers for tasks
Makes technical decisions
Designs software

220. Pick which 4 activities are the responsibilities of the Product Owner in Scrum.

Choice
Provides Estimates
Prioritize the backlog
Create User Stories
Commit to the Sprint
Perform user acceptance
Facilitate meetings
Champion Scrum
Perform release planning
Design software

221. Pick which 3 activities are the responsibilities of the Scrum Master in Scrum.

Choice
Provide Estimates
Prioritize the backlog
Commit to the Sprint
Perform user acceptance
Facilitate meetings

Choice
Champion Scrum
Volunteer for tasks
Make technical decisions
Remove impediments

222. What of the following is not a step in the Value Stream Mapping process?

Choice
Define the current state
Collect data
Amplify Learning
Depict the future state
Develop an implementation plan

223. At completion of iteration planning, the team has finished identifying the tasks they will commit to for the next iteration. Which of the following tools best provides transparency into the progress throughout the iteration?

Choice
Burndown chart
Gantt chart
Hours expended chart
Management baseline chart

224. A common reason a story may not be estimable is the:

Choice
Team lacks domain knowledge.
The story did not include a role.
Developers do not understand the tasks related to the story.
Team has no experience in estimating.

225. The purpose of a Sprint Retrospective is for the Scrum Team to:

Choice
Review stories planned for the next Sprint and provide estimates.
Demonstrate completed User Stories to the Product Owner.

Determine what to stop doing, start doing, and continue doing.
Individually provide status updates on the User Stories in progress.

226. Question: Which of the following BEST describes ROTI?

Choice
Measure of product backlog items (PBI) remaining
Measure of quality of features delivered in an iteration
Measure of required effort to complete an iteration
Measure of the effectiveness of the retrospective meeting

227. What Agile concept expresses delivering value in slices rather than in layers/stages?

Choice
Definition of Done
Value Mapping
Sashimi
Lean Value

228. In the Kano Model of customer satisfaction, this type of feature makes a product unique from its competitors and contributes 100% to positive customer satisfaction:

Choice
Excitement
Performance
Must-have
Threshold

229. Which of the following is NOT a principle from the Agile Manifesto?

Choice
Our highest priority is to satisfy the customer through early and continuous delivery of valuable software.
Business people and developers must work together daily throughout the project.
Continuous creation of technical debt and good design enhances agility.

> Working software is the primary measure of progress.

230. Which chart shows the total number of story points completed through the end of each iteration?

Choice
Iteration Burndown chart
Cumulative story point Burndown chart
Daily Burndown chart
Burnup chart

231. What is the order the hierarchy of product definition?

Order
User Story
Product Roadmap
Task
Epic
Product Vision
Theme

232. What Agile planning artifact is updated minimally once a year by the Product Owner?

Choice
Product Vision
Product Roadmap
Release Plan
Sprint Plan
Daily Plan

233. What Agile planning artifact should be updated at minimum semi-annually?

Choice
Product Vision
Product Roadmap
Release Plan

Sprint Plan
Daily Plan

234. What Agile planning artifact is created by the Product Owner and the development team?

Choice
Product Vision
Product Roadmap
Release Plan
Sprint Plan
Daily Plan

235. DSDM uses MoSCoW technique to create the prioritized requirements list. In MoSCoW technique, 'M' stands for:

Choice
Most useful
Must have
Must not have
Minimum marketable feature

236. Based upon this Burndown chart, is this project ahead of schedule or behind schedule?

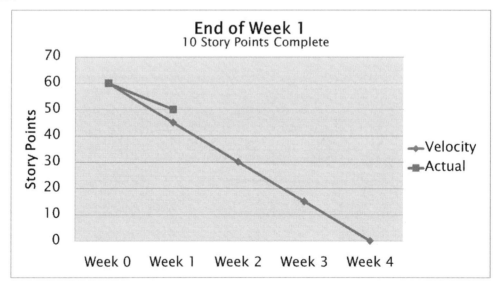

Choice
Ahead of schedule
Behind schedule

237. Which of the following occurs in the first Sprint?

Choice
Create a project plan
Develop a shippable piece of functionality
Complete your reference architecture
Develop the Product Roadmap

238. With multiple Scrum teams, you should have a separate product backlog.

Choice
True
False

239. What is the best definition of "Done"?

Choice
Whatever will please the Product Owner
It is determined by the Scrum Master
The product has passed QA and has all of the required release documentation
The definition of "Done" is one that would allow the development work to be ready for a release

240. Which of the following does NOT describe Scrum?

Choice
Simple to understand
A lightweight framework
Difficult to master
A process or a technique for building products

241. Scrum is NOT:

Choice
A set of software project management principles
Founded on empirical process control theory, or empiricism.
A process framework used to manage the development of products
Designed for static requirements

242. Which of the following is NOT part of the Scrum Framework?

Choice
Roles
Events
Characteristics
Artifacts

243. What are the three pillars of Scrum?

Choice
Transparency, Inspection, and Adaptation
Transparency, Inspection, and Empiricism
Transparency, Acceptance and Adaptation
Retrospectives, Inspection, and Adaptation

244. What does Scrum mean by Transparency?

Choice
Users can perform code reviews at any time
Documentation is available to anyone
All team members sit in a visible location
The process is understandable by all stakeholders

245. When does Adaptation occur in Scrum?

Choice
At the Sprint Review
During Sprint Planning
In the daily Scrum

Choice
At all four formal Scrum events
As Part of the Sprint Retrospective

246. Who is NOT part of the Scrum Team?

Choice
Product Owner
Scrum Master
Customer
Development Team

247. A cross-functional team in Scrum consists of which types of team members?

Choice
A specialist in QA
An architect
A release manager
Anyone with the skills to accomplish the work

248. Scrum is both an iterative and incremental Agile process.

Choice
True
False

249. When does Inspection occur?

Choice
Throughout the Sprint
Only at the end of the Sprint
Whenever the Product Owner wishes
Frequently, but not so often that it gets in the way of work

250. The Product Owner is the sole person responsible for managing the Product Backlog.

Choice
True

False

251. Who is responsible for maximizing the value of the product?

Choice
Senior Executives
The Product Owner
The Scrum Master
The Development Team

252. The Product Owner does not have to be a single person but may be a committee or a shared responsibility between multiple individuals.

Choice
True
False

253. No one, not even the Scrum Master, tells the development team how to build the product.

Choice
True
False

254. The development team should have a lead developer to ensure the work is properly executed.

Choice
True
False

255. The optimum size of the Scrum Team is:

Choice
7
Between 3 and 9
5
It depends

256. The Product Owner and Scrum Master are never part of the Development Team.

Choice
True
False

257. The Scrum Master as a Servant Leader is in service to which of the following?

Choice
The Development Team
The Organization
The Product Owner and the Development Team
The Organization, the Product Owner and the Development Team

258. Which one of the following is NOT a Scrum Event?

Choice
Sprint
Daily Scrum
Sprint Review
Weekly Status
They are all Scrum events.

259. A new Sprint starts immediately following the previous Sprint.

Choice
True
False

260. Sprints lengths can vary each Sprint as long as they don't exceed a month.

Choice
True
False

261. Put the following in order of first occurrence in a Sprint.

Order
Sprint Retrospective
Daily Scrum

Sprint Planning
Sprint Review

262. All of the following are true about change during a Sprint EXECPT:

Choice
Changes can be made that impact the Sprint goal
The development team can change tasks in the Sprint backlog
The Product Owner is the only person that can add or remove a User Story in the Sprint Backlog
Change may occur as scope is clarified between the Product Owner and the Development Team

263. The Scrum Team and Development Team are the same thing.

Choice
True
False

264. In Scrum, Sprints are never longer than a calendar month.

Choice
True
False

265. Who can cancel a Sprint?

Choice
The Development Team
Executive Stakeholders
The Product Owner
The Scrum Master

266. What happens if the customer no longer wants the feature that the Sprint Goal intended to meet?

Choice
The Development Team should determine if there is value in the Sprint

Choice
The Executive Stakeholders should determine if the Sprint should continue
The Product Owner should cancel the Sprint
The Scrum Master should cancel the Sprint

267. How long is the Sprint Planning meeting?

Choice
4 hours
8 hours
Depends on the length of the Sprint
3 hours

268. In the first part of the Sprint Planning meeting, what is NOT accomplished?

Choice
Items are selected from the Product Backlog
The Development Team decides how much work can be accomplished
The Scrum Team defines the Sprint Goal
The Tasks are defined

269. No one but the Scrum Team attends the Sprint Planning meeting.

Choice
True
False

270. Which three of the following points about the daily Scrum are TRUE?

Choice
It is time-boxed
It is held at the same place and time every day
The Product Owner provides an update
The Scrum Master enforces the rule that only Development Team members participate

271. Which of the following are TRUE about the Sprint Review? (Choose two).

Choice
It should be a formal presentation
Stakeholders may attend
The Product Owner presents what backlog items are "Done"
The Scrum Master demonstrates the product

272. The feedback from the Sprint Review impacts the next Sprint planning meeting.

Choice
True
False

273. Which of the following are TRUE about the Sprint Retrospective? (Choose 2)

Choice
It is three hours for a one month Sprint
It occurs before the Sprint Review
It is an opportunity to inspect the people, relationships, process, and tools in the last Sprint
It is the only time improvements are made during a Sprint

274. Match the following:

	Choice
Scrum Event	Product Backlog
Scrum Artifact	Gantt Chart
Scrum Role	Daily Scrum
Not Scrum	Product Owner

275. The Product Backlog is baselined at the start of the project and not changed for at least three Sprints.

Choice
True
False

276. Who is responsible for ordering the Product Backlog?

Choice
Senior Executives
The Product Owner
The Scrum Master
The Development Team

277. Which item is NOT an attribute of the Product Backlog?

Choice
Description
Order
Estimate
Value
Owner

278. Backlog Grooming and Backlog Refinement are the same thing.

Choice
True
False

279. Which of the following 2 statements are TRUE about Product Refinement?

Choice
Should take no more than 10% of the Development Team's time
The Scrum Master facilitates these sessions
Multiple Scrum Teams may participate in this process
The Product Owner is responsible for all estimates

280. Who tracks work remaining in the Product Backlog?

Choice
The Development Team
The Scrum Master
The Product Owner
Senior Executives

281. Who can change the Sprint Backlog during a Sprint?

Agile Certification Study Guide – Practice Questions

Choice
Senior Executives
The Product Owner
The Scrum Master
The Development Team

282. If there are multiple Scrum teams working on a product, each needs its own definition of Done.

Choice
True
False

283. To truly adopt Scrum, you must pick and choose what roles, artifacts, events, and rules are right for your organization.

Choice
True
False

Scrum exists only in its entirety and functions well as a container for other techniques, methodologies, and practices.

284. When was Scrum first introduced?

Choice
In 2001 with the Agile Manifesto
In 1995 at a conference presentation
At General Electric as part of a Lean approach
In 2000 along with Extreme Programming

285. Scrum is a container for other techniques and methodologies.

Choice
True
False

286. What type of process control is Scrum?

Agile Certification Study Guide – Practice Questions

Choice
Classical
Empirical
Inspection
Adaptive

287. The Scrum Master is a management role?

Choice
True
False

288. There must be a release every Sprint.

Choice
True
False

289. How often should Development Team members change?

Choice
No more than every three Sprints
Never
Each Sprint
As needed

290. What is a time-boxed event?

Choice
It happens at the same time as a conflicting task
It has a maximum duration
It has a minimum duration
It has a fixed place and time

291. When is a Sprint finished?

Choice
When the definition of "Done" is met

Agile Certification Study Guide – Practice Questions

When the Product Owner accepts the increment
When the time-boxed duration is met
When the work remaining is zero

292. Who updates work remaining during the Sprint?

Choice
Senior Executives
The Product Owner
The Scrum Master
The Development Team

293. Identify all members of a Scrum Team:

Choice
Customer
Stakeholder
Product Owner
Scrum Master
Project Manager
Development Team

294. Who is responsible for the Project Plan and Gantt Chart in Scrum?

Choice
Project Manager
Scrum Master
Product Owner
No Scrum role

295. How long is a Sprint Review?

Choice
2 hours
4 hours
It depends on the length of the Sprint

296. If the Sprint Backlog cannot be completed in a Sprint, who resolves the issue?

Agile Certification Study Guide – Practice Questions

Choice
Product Owner
Scrum Master
Development Team
Both the Product Owner and Development Team

297. If a customer really wants a feature added to a Sprint, how should the Development Team respond?

Choice
Add the feature into the current Sprint backlog
Escalate to the Scrum Master
Add the item to the Product Backlog for prioritization in the next Sprint
Ask the Product Owner to work with the customer

298. When does a Sprint get canceled or end early?

Choice
When the Sprint backlog is complete
When the Sprint Goal cannot be met
When the definition of "Done" is met
When a key resource is out sick

299. How long is the time-box for the daily Scrum?

Choice
It depends
5 minutes per person on the Development Team
15 minutes
Whatever the Team decides

300. How does the Scrum Master provide the most value to the Team?

Choice
By facilitating discussions between the Product Owner and the Development Team

Ensuring time-boxes are kept
Removing impediments to the Development Team
Scheduling Scrum events

301. Select the statements that are TRUE about the Product Owner. (Choose two)

Choice
The Product Owner can clarify the backlog during the Sprint
The Product Owner estimates the size of the Sprint backlog
The Product Owner prioritizes the Product backlog
The Product Owner defines the Sprint Goal before the Sprint Planning meeting

302. Who creates the Sprint Goal?

Choice
The Development Team
The Scrum Master
The Product Owner
The entire Scrum Team

303. In Scrum, the development team decides which events or ceremonies take place during a Sprint.

Choice
True
False

304. The Scrum Master is a participant in the Sprint Retrospective.

Choice
True
False

305. If the Development Team does not have all the skills to accomplish the Sprint Goal, the Scrum Master should:

Choice
Cancel the Sprint

Choice
Stop using Scrum
Have the development team determine the definition of "Done" and work through the Sprint backlog
Remove the impacted stories from the Sprint backlog

306. The Project Manager plays the following role in Scrum:

Choice
Collects the status from the Scrum Master
Updates the Burndown chart
Creates the release plan
There is no project manager role in Scrum

307. What is the purpose of the Sprint Review? (Choose three)

Choice
To collaborate with stakeholders
To inspect and adapt
To provide status on the Sprint
To demonstrate what is "Done"

308. Scrum dictates the use of User Stories.

Choice
True
False

309. Scrum is a software development methodology.

Choice
True
False

310. If the Development Team doesn't like the time of the daily Scrum, what should the Scrum Master do?

Choice
Find a time that is open on everyone's calendar
Let the Development Team come up with a new time

Ask the Team to try the existing time for one Sprint
Tell them that Scrum is immutable and that they need to stick to it

311. The backlog is ordered by:

Choice
The needs of the Product Owner
Risk
Complexity
Size

312. Who is responsible for maximizing the value of the product backlog?

Choice
The Customer
The Scrum Master and Product Owner
The Product Owner
The Development Team and Product Owner

313. What happens if all the necessary testing doesn't occur in a Sprint?

Choice
The User Story is moved to the next Sprint
Additional testers are added in the next Sprint
A risk of not creating a potentially shippable product occurs
The Burndown chart is updated

314. Pick roles that support the Scrum Master in removing impediments. (Choose two.)

Choice
The Development Team
Senior Management
The Product Owner
The Customer

315. Match each of the following items with its associated time-boxed duration for a one-month Sprint.

	Choice
Sprint Review	3 hours
Sprint Retrospective	4 hours
Sprint Planning	1 month
Sprint	15 minutes
Daily Scrum	8 hours

316. Match the activity (on the right) to the Scrum event (on the left).

	Choice
Sprint Planning	Adapt the definition of "Done"
Sprint Retrospective	Demonstrate Functionality
Daily Scrum	Sprint Goal creation
Sprint Review	Inspect and adapt

317. What Scrum event or artifact supports daily inspection and adaptation?

Choice
Product Backlog
Sprint Backlog
Sprint
Scrum
Working Product Increment

318. What Scrum event or artifact is the single source of requirements for any changes to be made to the product?

Choice
Product Backlog
Sprint Backlog
Sprint
Scrum
Working Product Increment

319. What Scrum event or artifact is the set of items selected for the Sprint, plus a plan for delivering the product Increment and realizing the Sprint Goal?

Choice
Product Backlog
Sprint Backlog
Sprint
Scrum
Working Product Increment

320. The Scrum Master's job is to work with the Scrum Team and the organization to increase the awareness of the artifacts. Which pillar of Scrum does this represent?

Choice
Transparency
Inspection
Adaptation

321. Scrum users must frequently review Scrum artifacts and progress toward a Sprint Goal to detect undesirable variances. Which pillar of Scrum does this represent?

Choice
Transparency
Inspection
Adaptation

322. Based upon this Burndown chart, is this project ahead of schedule or behind schedule?

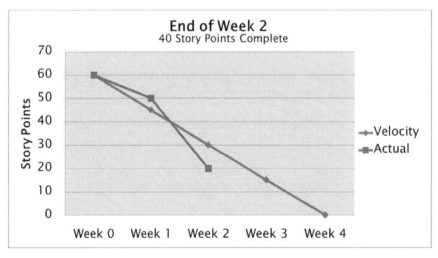

Agile Certification Study Guide – Practice Questions

Choice
Ahead of schedule
Behind schedule

323. Pick the two PMLC models that are based upon the Agile Project Management (APM) approach:

Choice
Linear
Adaptive
Incremental
Iterative

324. Which of the following is NOT a characteristic of an Adaptive PMLC Model?

Choice
Iterative Structure
Clear up front requirements
Mission Critical Projects
JIT Planning

325. This management approach is based on knowing well defined goals but not the means for a solution.

Choice
Traditional Project Management
Emertxe Project Management
Extreme Project Management
Agile Project Management

326. This Emertxe Project Management (MPx) approach is when neither a goal nor solution is clearly defined.

Choice
True
False

327. Every Project Management Life Cycle (PMLC) has a sequence of processes that include these phases:

> Scoping
> Planning
> Launching
> Monitoring & Controlling
> Closing

Choice
True
False

328. Which of the following is a weakness of an Adaptive PMLC Model?

Choice
Does not waste time on non-value-added work
Does not waste time planning uncertainty
Cannot identify what will be delivered at the end of the project
Avoids all management issues processing scope change requests

329. Sequence the core practices of Kanban in order of execution.

Order
Make process policies explicit
Implement Feedback Loops
Limit WIP
Visualize the workflow
Improve collaboratively
Manage the flow

330. The number of days needed between feature specification and production delivery is called:

Choice
Cycle time
Calendar time
Ideal time
Real time

331. The number of days needed between customer request and production delivery is called:

Choice
Cycle time
Lead time
Ideal time
Real time

332. Classes of Services in Kanban are used to:

Choice
Support estimation for Kanban Cards
Prioritize the queue by risk
All of the above
Ensure WIP limits are realistic

333. The purpose of Work in Progress (WIP) limits is to prevent the unintentional accumulation of work, so there isn't a bottleneck.

Choice
False
True

334. The following is a picture of which of the following Information Radiators?

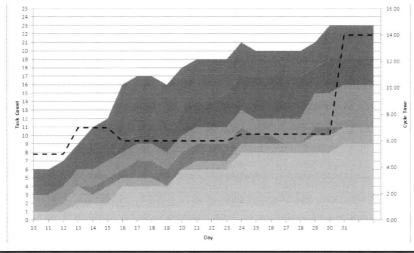

Choice
Burndown Chart
Kanban Tracking System

Cumulative Flow Diagram
Burnup Chart

335. The measure of productivity of a Kanban team is:

Choice
Cycle time
Lead Time
Work in Progress
Velocity
Throughput

336. Kanban cards should always be written using User Stories.

Choice
True
False

337. A term used to describe the work that can be delivered which meets the business requirements without exceeding them. (Choose Two)

Choice
Epic
Minimum Viable Product
Theme
User Story
Minimum Marketable Features

338. Order the 5 focusing steps of the Theory of Constraints.

Order
Subordinate Everything Else
Decide How to Exploit the Constraint
Identify the System Constraint
Elevate the Constraint
Go Back to Step 1, Repeat

339. This role champions the products, provides the budget and supports the Scrum Master in removing impediments

Choice
Subject Matter Expert
Product Owner
Business Owner
Project Manager

340. The behavior where the Scrum Team focuses on one or more stories until they are done is called:

Choice
Collaboration
Swarming
Pair-programming
Sprinting

341. Match the following roles on the right to the RASCI on the left:

	Choice
Accountable	Team Members
Consulted	Product Owner
Informed	Subject Matter Experts
Supportive	Stakeholders and Business Owner
Responsible	Scrum Master

342. The product owner should spend at least 3 hours per day with the development team?

Choice
True
False

343. Which role is external to the Scrum Team but provides a skills the does not exist on the Team?

Choice
Subject Matter Expert

Choice
Team Member
Scrum Master
Project Manager

344. Pick the 3 common Kanban Katas.

Choice
Daily Standup Meeting
Iteration Demo
Improvement Kata
Sprint Retrospective
Weekly Status
Operations Review

345. Which of the following is NOT considered an enterprise Agile method?

Choice
DAD
XP
LeSS
SAFe

346. Which Agile enterprise framework adopts and tailors methods such as Scrum, Extreme Programming (XP), and Agile Modeling (AM) in order to support scaling.

Choice
Disciplined Agile Delivery (DAD)
Agile Delivery Framework (ADF)
DSDM
Crystal

347. This Agile methodology's properties include Focus, Osmotic Communication and Project Safety.

Choice
Scrum
Crystal Clear

- 84 -

Agile Certification Study Guide – Practice Questions

Extreme Programming
Kanban

348. Which of the following frameworks has the following practices?

    Supports an envision, explore, adapt culture
    Supports a self-organizing, self-disciplined team
    Promotes reliability and consistency to the extent possible given the level of project uncertainty
    Provides management checkpoints for review

Choice
Dynamic Systems Development Method (DSDM)
Crystal Clear
Feature Driven Development (FDD)
Agile Delivery Framework

349. Which of the following is NOT a prioritization technique?

Choice
User Story Mapping
Kano analysis
Minimally Marketable Features (MMF)
Kitchen Prioritization
Planning Poker

350. Sequence the steps to User Story Mapping.

Order
User Story Decomposition
User Story Definition
Product Roadmap Definition
Release Planning

351. Choose which three statements are true about Approved Iterations:

Choice
Meets the definition of Done
The Architect has approved it

It is communicated to all Team members and stakeholders
As a result, the Product Owner updates Roadmaps and Release Plans
There is no technical debt

352. "Fail Sooner" is a benefit of Incremental Development.

Choice
True
False

353. In what order do you select requirements to work on in a risk adjusted backlog?

Order
Low Risk Low Value
High Risk High Value
High Risk Low Value
Low Risk High Value

354. Non-functional requirements should be written as user stories whenever possible.

Choice
True
False

355. CRUD is an acronym that defines a way to split stories. What does CRUD stand for?

Choice
Create/Update/Delete
Capture/Replace/Update/Define
Create/Review/Update/Done
Create/Replace/Update/Define

356. Match the story types on the left to their definition on the right:

	Choice
Architecturally Significant	Stories that figure out answers to tough technical or design problems
Analysis	Stories to improve the infrastructure the team is

	using
Infrastructure	Stories that are created to find other stories
Spike	Stories are functional Stories that cause the Team to make architectural decisions

357. Operations and Maintenance staff should not be part of the Agile team.

Choice
True
False

358. Which of the following is NOT a metric that measures the performance of DevOps in Agile?

Choice
Release date adherence percentage
Defects attributable to platform/support requirements
Percentage of NFRs met
Business value realized per release
Percentage increase in the number of releases

359. Which of the following is NOT a KPI (key performance indicator) of Agile?

Choice
Actual Stories Completed vs. Committed Stories
Quality Delivered to Customers
Team Enthusiasm
Team Attendance
Team Velocity
Technical Debt Management

360. Your project management office (PMO) has suggested your project could benefit from some self-assessment work at the next retrospective. Which of the following benefits would they most likely be looking to achieve from a self-assessment?

Choice
Assess compatibilities for pair programming assignments
Improve personal and team practices

Choice
Identify personal traits for human resources counseling
Gain insights for salary performance reviews

361. The definition of an Epic can be written using the acronym CURB. What does CURB stand for.

Choice
Create/Update/Replace/Big
Compound/Unknown/Risky/Basic
Complicated/Unusual/Really Big
Complex/Unknown/Risky/Big

362. The Japanese terms for an Agile developmental mastery model is:

Choice
Kaizen
Muda
Sashimi
Shu-Ha-Ri

363. Which development mastery model of skill acquisition lies in helping the teacher understand how to assist the learner in advancing to the next level?

Choice
Shu-Ha-Ri
Kaizen
Tuckman
Dreyfus

364. Which of the following is not part of Agile Discovery?

Choice
Document business outcomes that are quantifiable and measurable
Outline a plan for the technical and business architecture/design of the solution
Describe essential governance and organization aspects of the project and how the project will be managed
Define the tasks that the team will perform during an iteration

Agile Certification Study Guide – Practice Questions

365. Agile Analysis is a phase in the lifecycle of an Agile project.

Choice
False
True

366. Pick the THREE statements that are true about Agile Analysis.

Choice
It is a highly evolutionary and collaborative process
It occurs at the beginning and end of a project
It only includes the project team
It is communication rich
It explores the problem statement

367. It is not possible to have a fixed price contract in Agile.

Choice
True
False

368. Match the traditional contract model of the left with the Agile alternative on the right.

	Choice
Analysis, design, development and testing occur sequentially.	Change is accommodated within the non-contractual product backlog.
No value delivered until the entire project has been completed.	The highest risk and highest value items are tackled first.
There is no attempt to control the order in which the requirements are tackled.	Value delivered at the end of every Sprint.
Success is measured by reference to conformance with the plans.	There is concurrent design and development.
Changes 'controlled' by means of the change control mechanism.	Success is measured by reference to the realization of the desired business outcomes.

369. What is the Japanese business philosophy focused on making constant improvements?

Choice
Shu-Ha-Ri
Sashimi
Aikido
Kaizen

370. Which of the following is an Agile improvement technique to address issues continuously, e.g. after daily stand-up?

Choice
Retrospectives
Verification Sessions
Intraspectives
Futurespectives

371. Similar to inspect and adapt in Scrum, this can be represented as Build, Measure, Learn.

Choice
Six Sigma
Kaizen
DMAIC
Agile Learning Cycle

Agile Certification Questions with Answers

1. Which of the following is NOT a principle of the Agile Manifesto?

Correct	Choice
	Working software is delivered frequently
	Close, daily co-operation between business people and developers
X	Daily team meetings to review progress and address impediments
	Projects built around motivated individuals, who should be trusted

Additional feedback: The Agile Manifesto does not dictate meeting lengths or frequencies.

2. A key component of Agile software development is:

Correct	Choice
	Requirements should be complete before beginning development
	Change must be minimized
	Risk management is robust
X	Requirements are able to evolve during development

3. Which one is NOT a Pillar of Scrum?

Correct	Choice
	Transparency
	Inspection
	Adaption
X	Empiricism

Additional feedback: Transparency, Adaption, Inspection are the pillars of Scrum.

4. Which is NOT a Scrum Role?

Correct	Choice
	Product Owner
	Team Member
X	Project Manager
	Scrum Master

Additional feedback: Project Manager is not a Scrum Role

Agile Certification Study Guide – Answers

5. For a one month Sprint, what is the recommended duration of the Sprint Planning meeting?

Correct	Choice
X	2 sessions - 8 hours
	1 Session - 4 hours
	3 hours
	Decided by the Team

Additional feedback: Two sessions totaling 8 hours. 4 hours for the requirements session and 4 hours for the design session.

6. What is the definition of velocity?

Correct	Choice
	The number of Sprints per release
	The number of items resolved in a daily Scrum
X	The number of story points completed in Sprint
	The number of User Stories in a release

Additional feedback: Velocity is a metric that predicts how much work a team can successfully complete within a Sprint typically using Story Points.

7. In which meeting do you capture lessons learned?

Correct	Choice
	Sprint Planning
	Sprint Review
X	Sprint Retrospective
	Daily Status Meeting

Additional feedback: The Sprint Retrospective is an opportunity for the Scrum Team to inspect itself and create a plan for improvements to be enacted during the next Sprint.

8. Who developed extreme programming (XP)?

Correct	Choice
	Mike Cohn
	Ken Schwaber
X	Kent Beck
	Alistair Cockburn

Agile Certification Study Guide – Answers

Additional feedback: Extreme Programming (XP) is the name that Kent Beck has given to a lightweight development process he developed.

9. Which is NOT a role on an XP team?

Correct	Choice
	Coach
	Customer
X	Product Owner
	Programmer

Additional feedback: XP has their own terminology for roles. The similar roles in XP are the XP Customer and XP Tester.

10. In Pair Programming, one programmer is responsible for all the coding in an iteration then the programmers switch for the next iteration.

Correct	Choice
	True
X	False

Additional feedback: It is expected in pair programming that the programmers swap roles every few minutes or so.

11. Extreme Programming (XP) includes which of the following 3 practices?

Correct	Choice
X	Simple Design
X	Coding standards
	Mange WIP
X	Sustainable Pace
	Eliminate Waste

12. What is the role of the XP Coach?

Correct	Choice
	Defines the right product to build
	Determines the order to build
	Ensures the product works
X	Helps the team stay on process

Additional feedback: The XP Coach role helps a team stay on process and helps the team to learn.

Agile Certification Study Guide – Answers

13. Which 2 are goals of Lean Software Development?

Correct	Choice
X	Ensure value
	Simple Design
	Continuous integration
X	Minimize waste

Additional feedback: Ensure Value and Minimize Waste are both goals of Lean. Continuous integration and Simple design are core practices of XP.

14. Which is NOT a principle of Lean?

Correct	Choice
	Eliminate waste
X	Time-box events
	Deliver fast
	Delay commitment

Additional feedback: Time-boxed events are a practice of Scrum not Lean.

15. Kanban means _____ in Japanese?

Correct	Choice
	User Story
X	Signal card
	Visual card
	Production Line

16. Which is the first step in setting up Kanban?

Correct	Choice
	Place prioritized goals on the left column of the board
	Decide on limits for items in queue and work in progress
X	Map your current workflow
	Lay out a visual Kanban board

Additional feedback: The 5 steps of Kanban are: 1) Map your current workflow 2) Visualize your work 3) Focus on flow 4) Limit your work in process 5) Measure and improve

Agile Certification Study Guide –Answers

17. Which Agile framework adopts and tailors methods such as Scrum, Extreme Programming (XP), Agile Modeling (AM), Unified Process (UP), Kanban and Agile Data (AD) in order to support scaling.

Correct	Choice
	DSDM
	Crystal
X	Disciplined Agile Delivery (DAD)
	Agile Delivery Framework (ADF)

18. Project X has an IRR of 12%, and Project Y has an IRR of 10%. Which project should be chosen as a better investment for the organization?

Correct	Choice
	It depends on the payback period
	Project Y
X	Project X
	Project X or Y, depending on the NPV

Additional feedback: Project X: (IRR) internal rate of return is calculation to compare investment alternatives. A higher value for the internal rate of return is desired.

19. What is a product roadmap?

Correct	Choice
	A list of reports and screens
X	A view of release candidates
	Instructions for deployment
	A backlog prioritization scheme

Additional feedback: A product roadmap allows you to communicate where you want to take your product. You should be able to present at least three release candidates in your roadmap.

20. Your sponsor has asked for clarification on when releases of your product will ship and what those releases will contain. Which Agile deliverable would best answer their needs?

Correct	Choice
	Product demo
X	Product roadmap
	Product backlog

Agile Certification Study Guide – Answers

	Sprint backlog

Additional feedback: A product roadmap allows you to communicate where you want to take your product.

21. The acronym MoSCoW stands for a form of:

Correct	Choice
	Estimation
	Risk identification
X	Prioritization
	Reporting

Additional feedback: MoSCoW is a simple way to prioritize User Stories. MoSCoW stands for: Must have, Should have, Could have, Won't have.

22. What is the reason to develop personas as part of User Story creation?

Correct	Choice
	When the conversation is centered on the high-level flow of a process
X	When trying to better understand stakeholder demographics and general needs
	When trying to capture the high-level objective of a specific requirement
	When communicating what features will be included in the next release

Additional feedback: Personas are created to better understand stakeholder demographics and general needs.

23. When comparing communication styles, which of the following are true?

Correct	Choice
	Paper-based communication has the lowest efficiency and the highest richness.
	Face-to-face communication has the highest efficiency and the lowest richness.
	Paper-based communication has the highest efficiency and the lowest richness.
X	Face-to-face communication has the highest efficiency and the highest richness.

Agile Certification Study Guide – Answers

Additional feedback: Agile manifesto principle: The most efficient and effective method of conveying information to and within a development team is face-to-face conversation.

24. Highly visible project displays are called:

Correct	Choice
	Project radiators
	Information refrigerators
X	Information radiators
	Project distributors

Additional feedback: An information radiator is a large, highly visible display used by software development teams to track progress.

25. Well-written User Stories that follow the INVEST model include which attributes?

Correct	Choice
	Independent, Negotiable, Smart
	Valuable, Easy-to-use, Timely
X	Negotiable, Estimable, Small
	Independent, Valuable, Timely

Additional feedback: A well-written user story follows the INVEST model: Independent, Negotiable, Valuable, Estimable, Small, Testable.

26. Wireframe models help Agile teams:

Correct	Choice
	Test designs
X	Confirm designs
	Configure reports
	Track velocity

Additional feedback: Wireframes help to generate feedback, see potential problems with an interface and confirm designs.

27. Which of the following is the hierarchy of User Story creation?

Correct	Choice
	Task, User Story, Feature, Theme
X	Theme, Epic, User Story, Task
	User Story, Epic, Theme, Feature

Agile Certification Study Guide – Answers

> Goal, Epic, Activity, User Story

Additional feedback: A theme is the highest level of the story hierarchy. A Product Owner breaks down a theme into one or more epics. An epic is a group of related User Stories. User stories are then divided into one or more tasks.

28. Which of the following statements is true for measuring team velocity?

Correct	Choice
	Velocity is not accurate when there are meetings that cut into development time.
	Velocity measurements are disrupted when some project resources are part-time.
	Velocity tracking does not allow for scope changes during the project.
X	Velocity measurements account for work done and disruptions on the project.

29. Self-organizing teams are characterized by their ability to:

Correct	Choice
	Do their own filing
	Sit where they like
X	Make local decisions
	Make project-based decisions

Additional feedback: Team members are empowered with collective decision making and cross-functional skills, which increases their ability to self-organize and make local decisions.

30. High-performing teams feature which of the following sets of characteristics?

Correct	Choice
	Consensus-driven, empowered, low trust
	Self-organizing, plan-driven, empowered
X	Constructive disagreement, empowered, self-organizing
	Consensus-driven, empowered, plan-driven

31. Which of the following sets of tools is least likely to be utilized by an Agile team?

Correct	Choice
	Digital camera, task board

Agile Certification Study Guide – Answers

	Wiki, planning poker cards
X	WBS, PERT charts
	Smart board, card wall

Additional feedback: WBS and PERT charts are tools of traditional project management.

32. Who typically has the best insight into task execution?

Correct	Choice
	Project managers
X	Team members
	Scrum Masters
	Agile coaches

Additional feedback: Team members have the best insight into task execution.

33. A servant leadership role includes:

Correct	Choice
X	Shielding team members from interruptions
	Making commitments to stakeholders
	Determining which features to include in an iteration
	Assigning tasks to save time

Additional feedback: One of the ways that the Scrum Master, as a servant leader, protects the Team is by shielding it from interruptions.

34. All of the following are TRUE about communicating on distributed teams EXCEPT:

Correct	Choice
	Should consider instant messaging tools
X	Should have an easier Storming phase
	Need to spend more effort communicating
	Have a higher need for videoconferencing

Additional feedback: Distributed teams typically have a more difficult Storming phase

35. What is the sequence of Tuckman's stages of team formation and development progress?

Correct Order
Forming

Agile Certification Study Guide – Answers

Storming
Norming
Performing

36. The Agile Manifesto states we value some items over others. Match the items in the columns below so each item on the left is valued over the corresponding item on the right.

Correct	Choice
Individuals and Interactions	Processes and Tools
Working Software	Comprehensive Documentation
Customer Collaboration	Contract Negotiation
Responding to Change	Following a Plan

37. Your team is running three-week Sprints. How much time should you schedule for Sprint Review sessions?

Correct	Choice
	1 hour, 15 minutes
	45 minutes
X	3 hours
	6 hours

Additional feedback: 3 weeks. Allow one hour per week of Sprint.

38. A 4-hour Sprint Planning meeting is typical for a Sprint or Iteration that is how long?

Correct	Choice
	Four weeks
X	Two weeks
	One week
	Four days

Additional feedback: Two weeks. Sprint Planning is two hours per week of Sprint. One hour for Requirements Planning and one hour for Design.

39. In Agile project management, responding to change is valued over _____.

Correct	Choice
	Contract negotiation
X	Following a plan

Agile Certification Study Guide – Answers

	Customer collaboration
	Processes and tools

Additional feedback: Agile Manifesto Value: Responding to change over following a plan. That is, while there is value in the items on the right, we value the items on the left more.

40. The person responsible for the Scrum process, making sure it is used correctly and maximizing its benefit:

Correct	Choice
	Product Owner
	Coach
X	Scrum Master
	Scrum Manager

41. An artifact on an Agile project can best be described as:

Correct	Choice
X	A work output, typically a document, drawing, code, or model
	The deliverable from a Sprint retrospective
	The Agile model of persona
	A document that describes how work the work needs to be done

42. Product Backlog Items (PBI) described as emergent, are expected to:

Correct	Choice
	Become new technology requirements
X	Grow and change over time as users stories are completed
	Become the highest priority items
	Be the most recent stories added to the backlog

43. Which is a report of all the work that is "done?"

Correct	Choice
X	Burndown chart
	Completion chart
	Kanban chart
	Sashimi

Additional feedback: A Burndown chart shows the work that is done and the work remaining.

44. Traditional project management uses requirement decomposition. This can be comparable to _____ of Agile User Stories.

Correct	Choice
	Sashimi
	Definition of done
X	Disaggregation
	De-separation

Additional feedback: Disaggregation refers to splitting a story or feature into smaller, easier-to-estimate pieces.

45. In Agile, _____ is the primary measure of progress:

Correct	Choice
	Accelerated Burndown chart
	Reduced risk
	Increased customer satisfaction
X	Working software

Additional feedback: An Agile Manifesto Principle: Working software is the primary measure of progress.

46. User Stories are:

Correct	Choice
X	Negotiable
	Baselined and not allowed to change
	Created by the Agile Project Manager
	The foundation of the roadmap

Additional feedback: User Stories are not a contract. They are not meant to be precise, detailed specifications of a feature. They should always be Negotiable.

47. When maintaining the product backlog, this role represents the interests of the stakeholders, and ensures the value of the work completed:

Correct	Choice
	Scrum Master
	Agile Project Manager

Agile Certification Study Guide – Answers

Correct	Choice
X	Product Owner
	Sponsor

Additional feedback: The Product Owner represents the interests of the stakeholders, and ensures the value of the work completed.

48. Acronym describing the attributes of a product backlog:

Correct	Choice
X	DEEP
	INVEST
	ITERA
	DESPO

Additional feedback: DEEP: Detailed Appropriately, Estimated, Emergent, and Prioritized

49. Agile uses a number sequence for estimating. The series of numbers begin with 0, 1, 1 and are calculated by adding the previous two numbers to get the next number. This number sequence is called:

Correct	Choice
	Sashimi
	Velocity
	Capacity
X	Fibonacci

Additional feedback: The Fibonacci sequence begins with 0 and 1 and each subsequent number is the sum of the previous two. 0, 1, 1, 2, 3, 5, 8, 13, 21

50. This role would be responsible for determining when a release can occur:

Correct	Choice
X	Product Owner
	Scrum Master
	Development Team
	Agile Project Manager

Additional feedback: The Product Owner is responsible for release planning.

51. On an Agile team, the project leader works to remove impediments from blocking the team's progress. This is known as what type of leadership?

Agile Certification Study Guide – Answers

Correct	Choice
X	Servant
	Command and control
	Consensus-driven
	Functional management

52. Which should NOT take place at the daily Scrum?

Correct	Choice
X	The Product Owner gives an update
	The Scrum Master manages the time-box
	The Development Team answers the three questions
	Issues are raised and documented

Additional feedback: The Product Owner should not speak at the Daily Scrum unless they are working on a story in the Sprint backlog.

53. What is the purpose of practicing asking the "5 Why's"?

Correct	Choice
	To determine the scope of the Sprint
X	To determine the root cause of an issue
	To determine the end result
	To determine the prioritized backlog

Additional feedback: 5 Whys is an iterative question-asking technique used to explore the cause-and-effect relationships underlying a particular problem.

54. Which of the following describe the roles in pair programming?

Correct	Choice
	Pilot and the navigator
X	Driver and the navigator
	Coder and the planner
	Leader and the second chair

Additional feedback: The programmer at the keyboard is usually called the "driver", the other, also actively involved in the programming task but focusing more on overall direction is the "navigator."

Agile Certification Study Guide – Answers

55. Which Agile methodology runs one week iterations; leverages the use of pair programming; and includes the roles of coach, customer, programmer, tracker, and tester?

Correct	Choice
	Lean
	Agile One
	Scrum
X	XP

56. Continuous attention to _____ and good design enhances agility.

Correct	Choice
	Best architectures
X	Technical excellence
	Robust plans
	Change control

Additional feedback: Agile Manifesto Principle: Continuous attention to technical excellence and good design enhances agility.

57. The best description of a Sprint backlog is:

Correct	Choice
	Daily progress for a Sprint over the Sprint's length
	A prioritized list of tasks to be completed during the project
X	A prioritized list of requirements to be completed during the Sprint
	A prioritized list of requirements to be completed for a release

58. This Agile methodology focuses on efficiency and habitability as components of project safety.

Correct	Choice
	Scrum
	Kanban
	Extreme Programming
X	Crystal Clear

59. A time-boxed period to research a concept and/or create a simple prototype is called a(n):

Agile Certification Study Guide – Answers

Correct	Choice
	Sprint
	Iteration
X	Spike
	Retrospective

60. In Scrum, who is responsible for managing the team?

Correct	Choice
	Scrum Master
	Project Manager
X	Development Team
	Product Owner

Additional feedback: The Development Team is responsible for self-management.

61. Simplicity - the art of _____ - is essential

Correct	Choice
X	Maximizing the amount of work not done
	Minimizing the amount of work done
	Maximizing the customer collaboration
	Minimizing contract negotiation

Additional feedback: Agile Manifesto Principle: Simplicity-the art of maximizing the amount of work not done-is essential.

62. This approach includes a visual process management system and an approach to incremental, evolutionary process changes for organizations.

Correct	Choice
X	Kanban
	Scrum
	Extreme programming
	Agile Unified Process

63. Which following statement is the least accurate regarding the Burndown chart?

Correct	Choice
	It is calculated using hours or points

Agile Certification Study Guide – Answers

	It is updated by the development team daily
X	It provides insight into the quality of the product
	It reflects work remaining

Additional feedback: The Burndown chart does not measure quality. It measures work completed and work remaining.

64. What can be described as "one or two written sentences; a series of conversations about the desired functionality."

Correct	Choice
X	User Story
	Story point
	Epic
	Product roadmap

65. This artifact contains release names with expected dates and includes major features, client-side impacts, server-side applications, platform support and markets served:

Correct	Choice
	Risk Burndown graph
X	Product roadmap
	Release migration plan
	Product vision

66. Which statement is least accurate when providing a definition of "Done"?

Correct	Choice
	It is the exit criteria to determine whether a product backlog item is complete
	It may vary depending on the project
X	It is defined by the Scrum Master
	It becomes more complete over time

Additional feedback: The definition of "Done" is defined by the Development Team and the Product Owner, not the Scrum Master.

67. All of the following are among the seven principles of the Lean approach with the exception of:

Agile Certification Study Guide – Answers

Correct	Choice
	Amplified learning
	Decide as late as possible
	Build integrity in
X	Maximize the work performed

Additional feedback: Maximize the work performed is not a Lean approach.

68. Typically calculated in story points, this is the rate at which the team converts "Done" items in a single Sprint:

Correct	Choice
	Burndown rate
	Burn-up rate
X	Velocity
	Capacity

Additional feedback: Velocity is calculated in story points. It is the rate at which the team converts "Done" items in a single Sprint.

69. All of the following occur in the second half of the Sprint planning meeting EXCEPT:

Correct	Choice
X	The Development Team identifies improvements that it will implement in the next Sprint
	The Product Owner is answers questions and clarifies User Stories
	The Development Team commits to work in the Sprint
	Tasks are defined for the User Stories

70. The best architectures, requirements, and designs emerge from:

Correct	Choice
	Hand-picked teams
	Co-located teams
X	Self-organizing teams
	Cross-functional teams

Additional feedback: Agile Manifesto Principle: The best architectures, requirements, and designs emerge from self-organizing teams.

71. The Sprint retrospective:

Agile Certification Study Guide – Answers

Correct	Choice
X	Is intended to promote continuous process improvements
	Is held at the end of each release
	Is conducted to provide the sponsor with key information on team progress
	Is optional

72. An iteration prior to a release that includes final documentation, integration testing, training and some small tweaks is called:

Correct	Choice
X	Hardening Iteration
	Buffer Iteration
	Release Iteration
	Integration Iteration

Additional feedback: A hardening iteration is used for readying the product for production.

73. Analyzing the current organizational processes, per project requirements, and making needed process changes is called:

Correct	Choice
	Value Stream Mapping
	Release Planning
	Use Case Development
X	Process Tailoring

Additional feedback: Process tailoring takes an organization's standard process definition and tailors it to the specific needs of the project.

74. At the end of first iteration, the team finishes User Stories A, B and 50% of C. What is the team velocity?

> The story sizes were:
> Story A = 8 Points
> Story B = 1 Points
> Story C = 5 Points
> Story D = 3 Points

Correct	Choice
	11.5

Agile Certification Study Guide – Answers

X	9
	14
	16

Additional feedback: 9: Stories are either done or not done. There is no concept of an unfinished story or 50% complete.

75. Suppose 8 new members joined the development team, and the team size is now 15. The daily Scrum is getting noisy and exceeding the 15 minutes time-box. What is the most effective way to address this situation?

Correct	Choice
X	Divide the team into two teams with minimum dependency and have two separate daily Scrums.
	Do nothing; allow the large team to exceed the time-box by a few minutes each meeting.
	Increase the time-box for the daily Scrum to 30 minutes.
	Ask the team members to only update on the impediments and highlight only the important ones.

Additional feedback: The real problem is the team size. The best way to address it is to split the teams with minimal interdependencies and conduct two separate daily Scrums.

76. While developing a story during the iteration, team discovered new tasks that were not identified earlier. A newly discovered task is such that the User Story cannot be completed during the iterations. What is the most appropriate action for the team to perform?

Correct	Choice
X	Let the Product Owner decide if there is still a way to meet the iteration goals.
	Discuss the situation with the Scrum Master and see if there is still a way to meet the iteration goals.
	Drop the User Story and inform the Product Owner that it will be delivered in the next iteration.
	Modify the scope of other User Stories to allow completion of the Sprint backlog.

Additional feedback: Let the Product Owner decide if there is still a way to meet the Sprint goal. The Owner may choose to reduce the functionality of a story or drop one entirely.

77. Suppose your team velocity is 8 story points, and the product backlog items are ordered by priority as shown below. If you are in a Sprint Planning meeting and need to commit to the User Stories for the next iteration, which ones will you select?

Agile Certification Study Guide –Answers

Story 1 = 3 Story Points
Story 2 = 4 Story Points
Story 3 = 3 Story Points
Story 4 = 1 Story Points

Correct	Choice
	Story 1, 2, and 3
X	Story 1, 2, and 4
	Story 1 and 2
	Story 2, 3, and 4

Additional feedback: Stories 1, 2, 4: The team only commits to their velocity. They are allowed to take a lower priority story in order (4) to meet their target velocity,

78. During Sprint Planning, the architect is constantly estimating higher than everyone else, and team members are increasing their estimates to accommodate her. This is an example of:

Correct	Choice
	Dominating
	Influencing
	Dictating
X	Anchoring

Additional feedback: Anchoring is when people hear a number and even if that number is not relevant; it still seems to influence their estimates.

79. Traditional Project Management uses the Work Breakdown Structure to develop requirements in terms of activities. What type of breakdown structure is used in Agile for this purpose?

Correct	Choice
X	Feature Breakdown Structure
	Product Roadmap
	Product Breakdown Structure
	Sprint Backlog

Additional feedback: A Feature Breakdown Structure is an Agile artifact that displays the backlog of features for each release of the Agile project.

80. The Agile artifact that describes the incremental nature of how a product will be built and delivered over time, along with the important factors that drive each individual release is called:

Correct	Choice
	Product Vision Statement
	Product Wireframe
X	Product Roadmap
	Product Theme

81. Suppose you are performing integrated testing on each of the different product increments developed during an iteration to ensure that the increments work together as a whole. What type of iteration is this?

Correct	Choice
	Hybrid Iteration
X	Hardening Iteration
	Heuristic Iteration
	Handoff Iteration

82. Which of the following is an Agile practice promoted by XP that is often used in conjunction with other Agile methods?

Correct	Choice
	Dynamic Systems Development Method (DSDM)
	Feature Driven Development (FDD)
	Adaptive Software Development (ASD)
X	Test Driven Development (TDD)

Additional feedback: Test-driven development (TDD), also called test-driven design, is a method of software development in which unit testing is repeatedly done on source code.

83. All of the following are metrics used in Agile except:

Correct	Choice
X	Real Time
	Velocity
	Escaped defects
	Cycle time

Additional feedback: Velocity, Escaped Defects and Cycle Time are all metrics that are used in Agile Project Management. Real Time is an Agile estimation "sizing unit" that refers to the actual time during each day that the team members are available and are productively working on specific Agile project tasks.

Agile Certification Study Guide – Answers

84. Alistair Cockburn created the Crystal Family of Agile Methods, all of whose names contain colors of quartz crystals taken from geology. What two characteristics of an Agile project are used to determine the color of the Crystal method?

Correct	Choice
	Size and complexity
	Duration and complexity
X	Size and criticality
	Duration and criticality

Additional feedback: Crystal comes from Alistair Cockburn's characterization of projects along the two dimensions of "size" (meaning the size of the project team) and "criticality" (meaning the damage that will be caused if the developed product or system fails).

85. When implementing Agile Project Management, risk management that occurs simply by following Agile best practices such as iterative planning and review activities is called:

Correct	Choice
	Inherent risk management
X	Organic risk management
	Overt risk management
	Intrinsic risk management

86. One of the major tools and techniques used in Lean Software Development is Value Stream Mapping. What is the primary purpose of value stream mapping?

Correct	Choice
	To improve business processes
X	To identify and eliminate waste
	To ensure product quality
	To increase customer value

87. Which of the following is NOT recognized as a "unit" that can be used for estimating the size of the requirements on your Agile project?

Correct	Choice
	Real time
	Relative size
	Ideal time

Agile Certification Study Guide – Answers

Correct	Choice
X	Ideal size

Additional feedback: Real time, relative size and ideal time are all used in Agile estimating.

88. Extreme Programming (XP) defines four basic activities that are performed during the software development process. These include designing, coding, testing and ... ?

Correct	Choice
	Collaborating
	Leveling
	Communicating
X	Listening

Additional feedback: One of the four basic activities of XP is listening to the Customer and also to other development team members.

89. Agile project management and product development use several types of documents specific to each iteration; they are known as "artifacts." All of the following documents are Agile iteration artifacts except:

Correct	Choice
X	Iteration Vision Statement
	Iteration Backlog
	Iteration Plan
	Iteration Burndown Chart

Additional feedback: The Iteration Vision Statement is not an Iteration artifact. Product vision statements are created and updated annually.

90. In Sprint Planning, the development team estimates User Stories provided by the Product Owner and agrees to ones that can be completed in the Sprint. This is an example of:

Correct	Choice
	Osmotic Communication
X	Reciprocal Commitment
	Universal Acceptance
	Tacit Agreement

Additional feedback: This is an example of Reciprocal Commitment, where the development team commits to delivering the specified functionality according the definition of done and the Product Owner agrees not to change priorities during the Sprint.

Agile Certification Study Guide –Answers

91. Which of these statements is NOT correct about Ideal time and Calendar time?

Correct	Choice
	Ideal time is the time that is actually required to complete the work
X	Both of them convey the same meaning
	Calendar time is the amount of time that passes on clock (calendar days)
	Normally calendar days are not equal to ideal days

Additional feedback: Ideal time is not the same as Calendar time. Calendar is the available duration to complete work whereas Ideal time is the time required to complete the work.

92. Which of the following is NOT an Extreme Programming Practice?

Correct	Choice
	Pair Programming
	Planning Game
	Test Driven Development
X	Extreme Persona

Additional feedback: An extreme persona is a user modeling technique not specifically attributed to XP.

93. On XP teams, what is expected from the Project Manager?

Correct	Choice
	Coach the team on Agile practices
X	Help the team work with the rest of the organization
	Provide domain expertise to the team
	Define the software

94. Technical debt is the total amount of less-than-perfect _____ in your project.

Correct	Choice
	Defects
X	Design and implementation decisions
	Code commenting
	Code Sharing

95. _____ an estimate refers to estimating a story based on its relationship to one or more other stories.

Correct	Choice
X	Triangulating
	Triaging
	Aggregating
	Disaggregating

96. Which type of estimation refers to splitting a story into smaller and easier to manage pieces?

Correct	Choice
	Expert opinion
	Analogy
X	Disaggregation
	Bottom up

97. The _____ the technical debt means the _____ the intrinsic quality?

Correct	Choice
	higher, higher
X	higher, lower
	lower, lower
	lower, higher

Additional feedback: The higher the technical debt means the lower the intrinsic quality.

98. A technique in which a team collaboratively discusses acceptance criteria and then distills them into a set of concrete tests before development begins is called:

Correct	Choice
	Feature Driven Development (FDD)
X	Acceptance Test Driven Development (ATDD)
	Test Driven Development (TDD)
	User Story workshops

99. Setting up development work in a way that the team can figure out what to do next is called:

Correct	Choice
X	A pull system
	Push system

	Critical path
	Sprint backlog

Additional feedback: A pull system, used in Kanban, presents the development work in such a way that it clearly identifies what should be worked on next.

100. What Agile development approach is being used when the whole team works towards solving a complex problem?

Correct	Choice
X	Swarming
	Norming
	Performing
	Collaborating

Additional feedback: Swarming is when the whole team works towards solving a complex problem.

101. The acronym for a good product backlog is DEEP. What does it stand for?

Correct	Choice
	Detailed appropriately, Emergent, Estimated, and Practical
X	Detailed appropriately, Emergent, Estimated, and Prioritized
	Descriptive, Emergent, Estimated, and Prioritized
	Detailed appropriately, Exact, Estimated, and Prioritized

102. Which of the following is NOT a characteristic of an Agile plan?

Correct	Choice
	Follows rolling wave planning approach
	Are top down
	Easy to change
X	Shows dependencies of one task to others

Additional feedback: Showing dependencies of one task on others is not a characteristic of an Agile plan.

103. Which type of risk analysis does an Agile team use to identify risks on their project?

Correct	Choice
	Risk Burndown Chart
	Pareto Analysis

Correct	Choice
X	Qualitative Risk Analysis
	Quantitative Analysis

Additional feedback: Qualitative analysis uses judgment, intuition, and experience in determining risks and potential losses

104. The trend of work remaining across time in a Sprint, a release, or a product, with work remaining tracked on the vertical axis and the time periods tracked on the horizontal axis is called a _____.

Correct	Choice
X	Burndown Chart
	Burnup Chart
	Progress Chart
	Parking Lot Chart

Additional feedback: A Burndown Chart tracks work remaining on the vertical axis and the time periods on the horizontal axis

105. Which list below includes the attributes of a good User Story?

Correct	Choice
X	Small, estimable, independent, negotiable
	Testable, estimable, renewable, valuable
	Negotiable, small, explainable, valid
	Valuable, estimable, dependent, small

Additional feedback: INVEST in good User Stories. (Independent, Negotiable, Valuable, Estimable, Small, Testable)

106. The Agile Manifesto principle, "Our highest priority is to satisfy the customer through early and continuous delivery of valuable software," is achieved through which Scrum practice?

Correct	Choice
	Daily Scrum
X	Sprints
	Release Planning
	Sprint Planning

Additional feedback: Sprints which are no longer than one month and often shorter promote continuous delivery to the customer

Agile Certification Study Guide – Answers

107. During which meeting do team members synchronize their work and progress and report any impediments to the Scrum Master for removal?

Correct	Choice
	Sprint Planning meeting
X	Daily Scrum
	Sprint Retrospective
	Weekly Status meeting

108. Who is responsible for change management in Scrum projects?

Correct	Choice
	Project Manager
	Project Sponsor
	Scrum Master
X	Product Owner

109. The technique used to analyze the flow of information and materials through a system to eliminate waste is:

Correct	Choice
	Fishbone diagramming
	Flow charting
X	Value stream mapping
	Pareto analysis

Additional feedback: The value stream map is a Lean tool that practitioners use to analyze the value stream.

110. Which of the following defines the goal of testing in Lean software development?

Correct	Choice
X	Testing is to improve the process and quality
	Testing plays the most crucial role in ensuring the intrinsic value of the product
	In Lean software development, testing always refers to Test Driven Development
	Testing is primarily performed to ensure that we don't have escaped defects

Agile Certification Study Guide – Answers

Additional feedback: In Lean software development, testing is done to improve the process and quality

111. In Kanban, a diagram that describes the overall flow and provides a measurement for every significant step in the workflow is the:

Correct	Choice
	Kanban board
X	Cumulative Flow diagram
	Burn Down Chart
	Parking Lot Chart

112. A visual control used in Lean software development to represent the velocity of business solutions delivered over time is a:

Correct	Choice
X	Business Value Delivered Chart
	Burndown Chart
	Cumulative Flow Diagram
	Burn-Up chart

113. Suppose the current release will be complete after 6 two-week iterations. A team member is insisting that the Product Owner write the acceptance test cases for the entire release. How should the Scrum Master respond?

Correct	Choice
	Since the release duration is not very long, agree with the idea and explain the advantage of doing this to Product Owner.
	It's up to the team and Product Owner to determine if it is a good idea.
X	Explain that stories should only be discussed in detail prior to development. Elaborating stories not in the next couple of iterations is wasteful.
	Decide to dedicate first Sprint to elaborating all the stories

Additional feedback: Making decisions at the last possible moment is a key Agile concept. Detailed acceptance criteria should wait to be created until the iteration in which they will be needed.

114. Suppose your team is working to create a commercial website and is in the process of developing User Stories by role. One team member suggests rather than only thinking about the target user, we should also think of some exceptional users who use the system very differently. What type of user is this team member referring to?

Correct	Choice
	High Risk User
	Admin Role
X	Extreme Persona
	Performance Tester

Additional feedback: An extreme persona is a team-manufactured user that is strongly exaggerated in order to elicit requirements that are non-standard.

115. Which of the following testing is done manually by the XP team?

Correct	Choice
	Regression testing
	Unit testing
	Integration testing
X	Exploratory testing

Additional feedback: In XP, all testing tasks are automated. The only manual testing an XP team might perform is exploratory testing.

116. Which of the following are 2 attributes of Exploratory testing?

Correct	Choice
X	It involves minimum planning and maximum test execution
	It is typically automated
X	It is unscripted testing
	It is often the sole testing technique

Additional feedback: Exploratory testing is manual unscripted testing that has a minimum of planning but a high amount of execution. It is often done in conjunction with other testing.

117. A change made to the internal structure of software that makes it easier to understand and cheaper to modify but does not change its observable behavior is referred to as:

Correct	Choice
	A Spike
	Technical debt
	A User Story
X	Refactoring

Agile Certification Study Guide –Answers

Additional feedback: Refactoring is the process of changing a software system in such a way that it does not alter the external behavior of the code yet improves its internal structure

118. An iteration takes place in a time frame with specific start and end dates, called a time-box. Which of the following is NOT an advantage of time-boxing?

Correct	Choice
	Establishes a WIP limit
	Forces prioritization
	Demonstrates progress
X	Helps control technical debt

Additional feedback: Establishing a WIP limit, forcing prioritization, and demonstrating progress are all advantages of time-boxing.

119. Match each component of the Agile Triangle (on the left) to its associated description (on the right).

Correct	Choice
Value	Releasable Product
Quality	Reliable, Adaptable Product
Constraints	Cost, Schedule, Scope

120. Identify the three components of the Agile Triangle.

Correct	Choice
X	Quality
X	Value
	Cost
X	Constraints
	Scope
	Leadership

Additional feedback: Quality, Value, Constraints

121. There are four critical actions that should be embraced by an adaptive leader: improving speed-to-value, having a passion for quality, doing less, and _____.

Correct	Choice
X	Inspiring staff

Agile Certification Study Guide –Answers

	Managing conflict
	Facilitating meetings
	Ensuring effective communication

122. The 3 items are required for an Agile, adaptive environment:

Correct	Choice
X	People
X	Process
X	Product
	Tools
	Technology

Additional feedback: People, Process, Product are required for an Agile, adaptive environment.

123. What is the name of this facilitated process? One or more team members sequence the product backlog from smallest to largest User Story. The rest of the team validates the sequence. The whole team uses a sizing method such as T-shirt size or Fibonacci sequence to group the user stories.

Correct	Choice
	Relative estimation
	Pairwise comparison
	Planning Poker
X	Affinity estimating

124. While managing the Agile Product Lifecycle, match the frequency with which you update the five Agile plans.

Correct	Choice
Product Roadmap	Semi-Annually by The Product Owner
Sprint Plan	Each iteration by the team
Daily Plan (Scrum)	Daily by the individual
Release Plan	Quarterly by the Product Owner and teams
Product Vision	Annually by Product Owner

125. What are the 5 values of Agile Modeling?

Agile Certification Study Guide – Answers

Correct	Choice
X	Communication, Simplicity, Feedback, Courage, Humility
	Communication, Efficiency, Transparency, Courage, Humility
	Communication, Simplicity, Feedback, Adaptation, Continuous Improvement

126. Pick the one factor that is NOT part of the Agile Scaling Model.

Correct	Choice
	Team size
	Geographical distribution
	Domain complexity
X	Leadership Style
	Organizational distribution
	Technical complexity
	Organizational complexity

Additional feedback: Leadership Style is not part of Agile Scaling Model. The rest of them are.

127. What are the 5 phases of the Agile Project Management (APM) delivery framework?

Correct	Choice
X	Envision, Speculate, Explore, Adapt, Close
	Elaborate, Speculate, Examine, Adapt, Close
	Envision, Enable, Explore, Enhance, Close

128. The ultimate goal of _____ is to deploy all but the last few hours of work at any time.

Correct	Choice
X	Continuous Integration
	Collective Code Ownership
	Synchronous Builds
	Asynchronous Builds

129. When reading a burn-down chart, what does each status measurement say about project performance? Match the items below.

Agile Certification Study Guide –Answers

Correct	Choice
Actual Work Line is above the Ideal Work Line	Behind Schedule
Actual Work Line is below the Ideal Work Line	Ahead of Schedule
Actual Work Line is on the Ideal Work Line	On Schedule

130. How do you read a burndown bar chart? Match the phrases below to create instructions.

Correct	Choice
As tasks are completed	the top of the bar is lowered.
When tasks are added to the original set	the bottom of the bar is lowered.
When tasks are removed from the original set	the bottom of the bar is raised.
When the amount of work involved in a task changes	the top of the bar moves up or down.

131. What are the three components of an Agile Project Charter?

Correct	Choice
X	Success Criteria
X	Vision
	Objectives
	Scope
X	Mission
	Risks

132. Match each activity (on the left) to its definition (on the right).

Correct	Choice
Communication	A slide presentation by the Product Owner to stakeholders
Coordination	The developer submits code for testing. The UX designer checks that the

- 125 -

	developer implemented the elements correctly.
Cooperation	The Product Owner adjusts some story priority to meet the dependency of another team.
Collaboration	Pair programming

133. In XP, the practice that any developer can change any line of code to add functionality, fix bugs, improve designs, or refactor demonstrates:

Correct	Choice
X	Collective Code Ownership
	Source Code Control
	Pair Programming
	Continuous Integration

134. When the Agile team works in a single location, the team is said to be_____.

Correct	Choice
X	Co-located
	Distributed
	Outsourced
	Functional

135. Teams of members working in different physical locations are called:

Correct	Choice
	Co-located Teams
X	Distributed Teams
	Outsourced Teams
	Global Teams

136. On Agile teams, conflict is to be avoided at all cost.

Correct	Choice
	True
X	False

Additional feedback: Innovation occurs only with the free interchange of conflicting ideas, a phenomenon that was studied and documented by Hirotaka Takeuchi and Ikujiro Nonaka, the godfathers of Scrum.

Agile Certification Study Guide – Answers

137. Suppose you are a Scrum Master on a new Agile team. Which of the following strategies is best way to resolve conflict on the team?

Correct	Choice
X	Collaborate
	Negotiate
	Smooth over
	Ignore
	Use your authority

Additional feedback: Collaborate is a formal term for resolving conflict on a team.

138. Match the response options to each level of conflict.

Correct	Choice
Level 1: Problem to Solve	Collaboration or consensus
Level 2: Disagreement	Support and safety
Level 3: Contest	Accommodate, negotiate, get factual
Level 4 : Crusade	Establish safe structures again
Level 5 : World War	Do whatever is necessary

139. Which of the following is NOT one of the 5 common conflict types?

Correct	Choice
X	Compensation anxiety
	Lack of role clarity
	Difference in prioritizing tasks
	Working in silos
	Waiting on completion of task dependencies
	Lack of communication

Additional feedback: Compensation anxiety is not a conflict type. The rest of the list are the common types of conflict.

140. When we use the term "container" in Scrum what are we referring to?

Correct	Choice
X	A Sprint or Iteration
	Source code repository
	Development team room

Agile Certification Study Guide – Answers

A vertical slice of functionality

Additional feedback: A container is a closed space where things can get done, regardless of the overall complexity of the problem. In the case of Scrum, a container is a Sprint, an iteration.

141. Prioritize from high to low the sequence of User Story development.

Correct Order
High-value, high-risk stories
High-value, low-risk
Lower-value, low-risk
Low-value, high-risk

142. The number of days needed between feature specification and production delivery is called:

Correct	Choice
X	Cycle time
	Real time
	Ideal time
	Calendar time

Additional feedback: Cycle time is the number of days needed between feature specification and production delivery.

143. The PM Declaration of Interdependence is a set of six management principles initially intended for project managers of Agile software development projects. Match the items below to identify the principles.

Correct	Choice
We deliver reliable results	by engaging customers in frequent interactions and shared ownership
We expect uncertainty	and manage for it through iterations, anticipation, and adaptation
We boost performance	through group accountability for results and shared responsibility for team effectiveness
We improve effectiveness and reliability	through situationally specific strategies, processes and practices
We unleash creativity and innovation	by recognizing that individuals are the ultimate source of value.

| We increase return on investment | by making continuous flow of value our focus |

144. All of the following are attributes of the definition of "Done", EXCEPT:

Correct	Choice
X	It is a static artifact
	It is an audible checklist
	It is a primary reporting mechanism for team members on User Story progress
	It is crucial to a high-performing team

Additional feedback: The Definition of Done is not static. It should be improved over time.

145. The way that we calculate the number of years it takes to break even from undertaking a project which also takes into account the time value of money is the:

Correct	Choice
	Pay-back period
X	Discounted pay-back period
	NPV
	Cumulative cash flow

Additional feedback: In contrast to an NPV analysis, which provides the overall value of a project, a discounted payback period gives the number of years it takes to break even from undertaking the initial expenditure.

146. DRY is an acronym for which Agile development principle?

Correct	Choice
	Development Requires You
X	Don't Repeat Yourself
	Deploy RepeatedlY
	Develop, Refactor, Yagni

Additional feedback: The DRY (Don't Repeat Yourself) Principle states: Every piece of knowledge must have a single, unambiguous, authoritative representation within a system.

147. When is the best time to perform Earned Value Measurement (EVM) in Agile projects?

Agile Certification Study Guide – Answers

Correct	Choice
X	After the iteration
	After a release
	During an iteration
	Never - we don't perform EVM in Agile

Additional feedback: Typically, in Agile EVM, a measure of performance on an Agile project is performed after the iteration.

148. Emotional intelligence includes all of the following except:

Correct	Choice
	Self-awareness
	Motivation
X	Commitment
	Influence
	Intuitiveness
	Conscientiousness

Additional feedback: The 7 Elements of Emotional Awareness are: Self Awareness, Emotional Resilience, Interpersonal Sensitivity, Motivation, Influence, Intuitiveness, and Conscientiousness

149. In a burndown chart, if the remaining work line is above the expected work line, what does this signify?

Correct	Choice
	The project is ahead of schedule
X	The project is behind schedule
	The resources are performing above expectation
	The project is being well managed

150. Empirical process control constitutes a continuous cycle of inspecting the process for correct operation and results and adapting the process as needed. What characteristics does this apply to in Scrum?

Correct	Choice
	Self-organization, Collaboration and Time-boxing
	Quality, Cost and Scope
	Scrums, Sprint and Releases

Agile Certification Study Guide – Answers

X	Transparency, Inspection and Adaptation

151. Bugs reported by the customer that have slipped by all software quality processes are represented in this metric.

Correct	Choice
	Technical debt
X	Escaped defects
	Risk burndown
	Code quality

152. Testing that often occurs between "Done" and "Done, Done" is:

Correct	Choice
X	Exploratory testing
	Acceptance testing
	Unit testing
	Test driven development

Additional feedback: Exploratory testing, unscripted manual testing, is typically done after all of the development iteration are complete and are being made ready for production, sometimes called "Done, Done"

153. Which of the following is NOT one of the 12 core practices of XP:

Correct	Choice
	The Planning Game
X	Planning Poker
	Small Releases
	System Metaphor

Additional feedback: Planning Poker is an Agile estimating technique

154. Which of the following is NOT one of the 12 core practices of XP?

Correct	Choice
	Simple Design
	Continuous Testing
X	Vertical Slicing
	Refactoring

Agile Certification Study Guide – Answers

Additional feedback: Vertical slicing is taking a backlog item that has a database component, some business logic and a user interface and breaking it down into small stepwise progressions where each step cuts through every slice.

155. Which of the following is NOT one of the 12 core practices of XP?

Correct	Choice
	Pair Programming
	Collective Code Ownership
	40-Hour Work Week
X	Minimize Waste

Additional feedback: Minimize Waste is a Lean principle.

156. Which of the following is NOT one of the 12 core practices of XP?

Correct	Choice
X	Visualize the flow
	On-site Customer
	Coding Standards
	System Metaphor

Additional feedback: Visualize the flow is a Kanban Practice

157. Which of the following is NOT a reason to use a Feature Breakdown Structure (FBS) instead of a Work Breakdown Structure (WBS)?

Correct	Choice
	It allows communication between the customer and the development team in terms both can understand.
X	It allows you to baseline your project plan due to absence of change.
	It allows tracking of work against the actual business value produced.
	It allows the customer to prioritize the team's work based on business value.

Additional feedback: A Feature Breakdown Structure does not allow you to baseline your project plan due to absence of change. That is a use of the Work Breakdown Structure in traditional project management.

158. Team members who are part-time on your project will see at least a 15% reduction in their productivity per hour. The type of resource model in Agile is called:

Agile Certification Study Guide – Answers

Correct	Choice
	Collocated
X	Fractional assignments
	Distributed resources
	Over-allocated resources

Additional feedback: Fractional Assignments impact team productivity. Team members who are part-time on your project will see at least a 15% reduction in their productivity per hour.

159. Frequent verification and validation is key in Agile but each approach produces a very different result. Verification determines _____ whereas validation determines _____.

Correct	Choice
	if the product is "done" | if the product is "done - done"
X	if I am I building the product right | if I am I building the right product
	if I am I building the right product | if I am I building the product right
	if the product has passed unit testing | if the product has passed acceptance testing

160. What type of time estimation excludes non-programming time?

Correct	Choice
X	Ideal Time
	Calendar time
	Duration
	Real Time

Additional feedback: Ideal Time excludes non-programming time. When a team uses Ideal Time for estimating, they are referring explicitly to only the programmer time required to get a feature or task done, compared to other features or tasks.

161. Which of the following is an example of an information radiator?

Correct	Choice
	An email of a status report
	A text of a quick question to the Product Owner
X	A whiteboard showing the state of work
	A face-to-face conversation

Agile Certification Study Guide – Answers

Additional feedback: Whiteboards showing state of work. An information radiator is a large, highly visible display used by software development teams to track progress.

162. Assuming all projects require the same amount of up-front investment, the project with the highest _____ would be considered the best and undertaken first.

Correct	Choice
	Earned Value Management (EVM)
X	Internal Rate of Return (IRR)
	Net Present Value (NPV)
	Budget at Completion (BAC)

Additional feedback: Generally speaking, the higher a project's internal rate of return (IRR), the more desirable it is to undertake the project.

163. Match each Agile requirement type (on the left) to its definition (on the right).

Correct	Choice
Feature	Business solution, capability or enhancement that ultimately provides value to the business.
User Story	Describes the interaction of the users with the system
Story	Any requirement that is NOT a User Story (e.g. technical enabling, analysis, reminder to have conversation)
Task	Fundamental unit of work that must be completed to make a progress on a Story

164. 80% of the value comes from 20% of the work. Which law is this referring to?

Correct	Choice
	Parkinson's Law
	Moore's Law
X	Pareto's Law
	Jevon's Paradox

Additional feedback: Pareto's law is more commonly known as the 80/20 rule. This means that typically 80% of your results come from only 20% of your efforts.

165. Based on the following information, determine the number weeks until the next release.

   Length of a Sprint = 2 weeks
   Velocity of team = 35 points
   Number of story points assigned to minimum marketable features (MMF) = 280 points

Agile Certification Study Guide –Answers

Correct	Choice
	8 weeks
	12 weeks
X	16 weeks
	9 weeks

Additional feedback: 16 weeks. 280 points divided by 35 points of velocity = 8 Sprints. 8 Sprints of two weeks = 16 weeks.

166. What is the correct sequence of activities in release planning?

Correct Order
Identify features
Prioritize features
Split features using the MMF perspective
Estimate the value of the features
Estimate the cost of the features
Write stories for features
Create release plan by date or scope

167. Sequence the following concepts to create the popular acronym for creating good User Stories.

Correct Order
Independent
Negotiable
Valuable
Estimable
Small
Testable

168. Match the definitions (on the right) to each of the characteristics of a good User Story (on the left).

Correct	Choice
Independent	Stories can be worked on in any order
Negotiable	A story is not a contract
Valuable	If a story does not have discernable

Agile Certification Study Guide – Answers

	value it should not be done
Estimable	A story has to be able to be sized so it can be properly prioritized
Small	User Stories average 3-4 days of work
Testable	Each story needs to be proven that it is "done"

169. At minimum, all Kanban boards should have the following columns:

Correct	Choice
	To-Do, Doing, Done
	Analysis, Design, Develop, Test, Deploy
	Backlog, Design, Develop, Unit Test, Acceptance Test, Ready-to-ship
X	The Kanban columns are determined by the team

170. The Kano Model supports what Agile planning activity?

Correct	Choice
	Estimation
X	Prioritization
	Sizing
	Continuous Integration

Additional feedback: The Kano model is an effective technique to prioritize the product backlog by customer satisfaction.

171. Which one is NOT a level of need in the Kano Model?

Correct	Choice
	Basic Needs
	Performance Needs
X	Enabling Needs
	Excitement Needs

Additional feedback: Different categories of customer satisfaction in the Kano model include basic needs, performance needs, and excitement needs.

172. Match a definition (on the right) to a Lean principle (on the left).

Correct	Choice
Eliminate Waste	Create nothing but value

Create Knowledge	Maintain a culture of constant improvement
Build Quality In	Refactor - eliminate code duplication to zero
Defer Commitment	Schedule irreversible decisions at the last responsible moment
Optimize the Whole	Focus on the entire value stream
Deliver Fast	Limit work to capacity

173. Net present value (NPV) is a ratio that compares the value of a dollar today to the value of that same dollar in the future. An NPV that is negative suggests what?

Correct	Choice
	The project should be rejected
	I don't have enough information
X	The project should be deferred
	The project should be put on hold until the value is 0

Additional feedback: A negative NPV suggests that the project should be deferred over another project that has a positive NPV.

174. What is the Agile Open Space concept?

Correct	Choice
	When cubicles walls are removed for an Agile team.
X	It is a meeting designed to allow Agile practitioners to meet in self-organizing groups where they can share their latest ideas and challenges.
	The choice to collocate all team members for the beginning of a project.
	It is a core principle of ADSM

175. Osmotic communication is when team members obtain information from overheard conversations.

Correct	Choice
X	True
	False

Additional feedback: Osmotic communication means that information flows into the background hearing of members of the team, so that they pick up relevant information as though by osmosis.

Agile Certification Study Guide – Answers

176. The 80/20 rule is also known as what law?

Correct	Choice
	Little Law
X	Pareto's Law
	Mohr's Law
	The Law of Averages

Additional feedback: The Pareto principle (also known as the 80-20 rule, the law of the vital few, and the principle of factor sparsity) states that, for many events, roughly 80% of the effects come from 20% of the causes.

177. Pareto Analysis is the exercise of determining what 40% of functionality can be delivered with 60% of the effort.

Correct	Choice
	True
X	False

Additional feedback: Pareto Analysis is the exercise of determining what 80% of functionality can be delivered with 20% of the effort.

178. The length of time to recover the cost of a project investment is the:

Correct	Choice
	Net Present Value
X	Payback Period
	Earned Value
	ROI

Additional feedback: Payback Period

179. An archetypal user of a systems is called a(n):

Correct	Choice
	Super user
	Admin
X	Persona
	UX engineer

180. Personas are used in Agile requirements to depict which type of user?

Agile Certification Study Guide – Answers

Correct	Choice
	Real users
X	Fictitious users
	Super users
	Beta testers

181. When is Planning Poker used?

Correct	Choice
	During backlog prioritization
	As part of Pareto Analysis
X	During User Story sizing and estimating
	As part of the Daily Stand-up

Additional feedback: Planning Poker® is a consensus-based estimating technique. Planning Poker can be used with story points, ideal days, or any other estimating unit.

182. Process Tailoring is the iterative approach implementing your SDLC process.

Correct	Choice
X	True
	False

Additional feedback: Process tailoring is best done in an iterative manner: tailor some, implement some, and then repeat.

183. Product roadmaps are more accurate the closer we get to an actual release.

Correct	Choice
X	True
	False

Additional feedback: Product roadmaps should be updated and made more accurate the closer we get to an actual release.

184. Which is the process of continuously improving and detailing a plan as more detailed and specific information and more accurate estimates become available as the project progresses?

Correct	Choice
	Process Tailoring

	Pareto Analysis
X	Progressive Elaboration
	Open Space Planning

185. What is a change made to the internal structure of software to make it easier to understand and cheaper to modify without changing its observable behavior?

Correct	Choice
	Pair Programming
	Continuous Improvement
	Test Driven Development
X	Refactoring

186. Refactoring is a key way of preventing technical debt.

Correct	Choice
X	True
	False

Additional feedback: The key to managing technical debt is to be constantly vigilant, avoid using shortcuts, use simple design, and refactor relentlessly.

187. Which Agile estimation technique is based upon relative sizing?

Correct	Choice
	Ideal time
	Bottom up
X	Story points
	Little's Law

Additional feedback: Story points are relative values that do not translate directly into a specific number of hours.

188. What kind of User Story is written to provide an opportunity to research a solution in order to provide an estimate?

Correct	Choice
	Sprint Story
	Persona Story
X	Spike Story
	Needle Story

Agile Certification Study Guide – Answers

189. Sequence the activities that occur in a Retrospective meeting.

Correct Order
Set the Stage
Gather Data
Generate Insights
Decide What to Do
Close the Retrospective

190. Triple Nickels is a technique used in what kind of meeting?

Correct	Choice
	Sprint Planning
	Daily Scrum
X	Sprint Retrospective
	XP Planning Game

Additional feedback: Triple Nickels is a Sprint Retrospective technique. It is a brainstorming approach where it is very easy to have everyone express their thoughts which encourage everybody to be attentive to what others think.

191. Who is responsible for managing ROI in Agile projects?

Correct	Choice
	The Project Sponsor
X	The Product Owner
	The Agile Project Manager
	The Scrum of Scrums Master

Additional feedback: In Scrum, the product owner has full product responsibility, which includes securing funding, managing to ROI objectives, release planning and more.

192. The purpose of the Scrum of Scrums is to perform what function?

Correct	Choice
	To increase knowledge of Agile within the organization
	To provide dashboard reporting to executives
X	To manage cross-team dependencies working on the same project or product
	To ensure team building and staff development occurs

Agile Certification Study Guide –Answers

Additional feedback: When you scale Scrum to more teams, you handle dependencies and coordination among teams working on the product with a Scrum of Scrums

193. Which one is NOT one of the 5 common risk areas mitigated by Agile.

Correct	Choice
	Intrinsic schedule flaw
	Specification breakdown
	Scope creep
X	Stakeholder apathy
	Personnel loss
	Productivity variation

Additional feedback: The book *Waltzing with Bears* lists the 5 common risk areas as: 1) Intrinsic Schedule Flaw 2) Specification Breakdown 3) Scope Creep 4) Personnel Loss 5) Productivity Variance

194. In what order do your select requirements to work on in a risk adjusted backlog?

Correct Order
High Risk High Value
Low Risk High Value
Low Risk Low Value
High Risk Low Value

Additional feedback: The order is: 1) High Risk, High Value 2) Low Risk, High Value 3) Low Risk, Low Value 4) High Risk, Low Value

195. Which Agile method promotes the practice of risk-based Spike or Spike solutions?

Correct	Choice
	Scrum
	AgileUP
	ADSM
X	Extreme Programming

Additional feedback: Spikes, another invention of XP are a special type of story used to drive out risk and uncertainty in a user story or other project facet.

196. Which one is NOT a reason to perform a Spike?

Correct	Choice
	To perform basic research to familiarize the team with a new

Agile Certification Study Guide – Answers

	technology or domain
	To analyze the expected behavior of a large story so the team can split the story into estimable pieces.
X	To defer a story until a later Sprint while still showing progress to the Product Owner
	To do some prototyping to gain confidence in a technological approach

Additional feedback: Spikes are used for different reasons: perform research, analyze a story or do prototyping to validate a design approach, not to defer the work.

197. Which artifact is useful for seeing total project risk increasing or decreasing over time?

Correct	Choice
	Burndown bar chart
	Risk Burn-Up chart
X	Risk Burndown Graph
	Risk Map

198. On a risk map or a risk heat map, the vertical and horizontal axes represent:

Correct	Choice
	Effort and Impact
X	Probability and Impact
	Probability and Exposure
	Impact and Exposure

199. The Project Leader's primary responsibilities are to "move boulders and carry water." What is this an example of?

Correct	Choice
X	Servant leadership
	Leadership by example
	Command and control leadership
	The leadership metaphor

Additional feedback: As a servant leader, the Scrum Master's primary responsibilities are to "move boulders and carry water"-in other words, remove obstacles that prevent the team from delivering business value, and to make sure the team has the environment they need to succeed.

Agile Certification Study Guide – Answers

200. In XP, what is the practice of creating a story about a future system that everyone - customers, programmers, and managers - can tell about how the system works?

Correct	Choice
	Extreme persona
	Wireframe
X	System metaphor
	Simple design

Additional feedback: Kent Beck, author of *Extreme Programming Explained* defines a system metaphor as: "a story that everyone - customers, programmers, and managers - can tell about how the system works."

201. What Agile requirements management approach displays a roadmap using the following approach?

> The horizontal axis shows a high level overview of the system under development and the value it adds to the users.
>
> The vertical axis organizes detailed stories into releases according to importance, priority, etc.

Correct	Choice
	Release Planning Matrix
X	User Story Map
	Agile Requirements Map
	User Story Burndown Map

Additional feedback: The User Story Map presents the product roadmap with features on the horizontal axis and detailed stories ordered by priority on the vertical axis.

202. Which XP practice promotes the restriction on overtime?

Correct	Choice
X	Sustainable Pace
	Pair Programming
	Servant Leadership
	Small Releases

Additional feedback: Sustainable pace suggests working no more than 40 hours a week, and never working overtime a second week in a row.

203. What is the Agile term for the time period when some or all of the following occur: beta testing, regression testing, product integration, integration testing, documentation, defect fixing.

Agile Certification Study Guide – Answers

Correct	Choice
	Spike
	Code Freeze
X	Tail
	Lag

204. Agile development prevents technical debt.

Correct	Choice
	True
X	False

Additional feedback: Agile development does not prevent technical debt. Technical debt is often more prevalent in Agile due to the urgency to "ship" the software.

205. In Agile development, what is the term for the internal things that you choose not to do now, knowing they will impede future development if left undone?

Correct	Choice
	Escaped defects
	Verification and validation results
X	Technical debt
	Intrinsic quality

Additional feedback: Technical Debt is the development that you choose not to do now, knowing that it will impede future development if left undone.

206. What is the purpose of running a test before you develop the code?

Correct	Choice
	To complete all test cases
X	To ensure it fails
	To ensure it passes
	To be cross-functional

Additional feedback: You run a test first to ensure it fails. If the test passes before you have written the code then it is a flawed test.

207. Match the time box to the Scrum meeting for a one-month Sprint.

Agile Certification Study Guide – Answers

Correct	Choice
Daily Scrum	15 Minutes
Sprint Review	4 hours
Sprint Planning	8 Hours
Sprint Retrospective	3 hours

208. A reminder for the developer and Product Owner to have a conversation is:

Correct	Choice
	The Sprint planning meeting
	Backlog grooming
X	A User Story
	An Agile reminder

209. Wideband Delphi is used by an Agile Project manager to support what activity?

Correct	Choice
	Prioritization
	Scheduling
X	Estimation
	Risk Management

Additional feedback: The Wideband Delphi estimation method is a consensus-based technique for estimating effort.

210. The purpose of Work in Progress (WIP) limits is to prevent the unintentional accumulation of work, so there isn't a bottleneck.

Correct	Choice
X	True
	False

Additional feedback: One of the core properties in the Kanban method is that Work in Progress is limited. Limiting WIP is to match team's development capacity and to prevent bottlenecks.

211. Which 5 roles are defined by Extreme Programming?

Correct	Choice
	Scrum Master
X	Coach

Agile Certification Study Guide – Answers

X	Customer
	Stakeholder
X	Programmer
X	Tracker
	Product Owner
X	Tester

212. Simple Design, Pair Programming, Test-Driven Development, Design Improvement are all practices of which Agile methodology?

Correct	Choice
	Scrum
	Feature Driven Development (FDD)
X	Extreme Programming (XP)
	Dynamic Systems Development Method (DSDM)
	Crystal Clear
	Rational Unified Process (RUP)
	Agile Unified Process (AgileUP)

213. Which of the following is not an Agile methodology?

Correct	Choice
	Scrum
	Feature Driven Development (FDD)
	Extreme Programming (XP)
	Dynamic Systems Development Method (DSDM)
X	Program Evaluation Review Technique (PERT)
	Crystal Clear
	Rational Unified Process (RUP)
	Agile Unified Process (AgileUP)

Additional feedback: Program Evaluation Review Technique (PERT) is a traditional project management scheduling technique. All of the others are Agile techniques.

214. Incremental delivery means:

Correct	Choice
	Deliver nonfunctional increments in the iteration retrospectives.

	Release working software only after testing each increment.
	Improve and elaborate our Agile process with each increment delivered.
X	Deploy functional increments over the course of the project.

215. When we practice active listening, what are the levels through which our listening skills progress?

Correct	Choice
	Global listening, Focused listening, Intuitive listening
	Interested listening, Focused listening, Global listening
	Self-centered listening, Focused listening, Intuitive listening
X	Internal listening, Focused listening, Global listening

216. Which Agile method goes through the following stages:

    1a. The Feasibility Study
    1b. The Business Study
    2. Functional Model Iteration
    3. System Design and Build Iteration
    4. Implementation

Correct	Choice
	Rational Unified Process (RUP)
	Feature Driven Development (FDD)
X	Dynamic Systems Development Method (DSDM)
	Lean Software Development
	Scrum

217. What is a Japanese term used in Lean software development is an activity that is wasteful, unproductive, and doesn't add value?

Correct	Choice
	Sashimi
	Kanban
X	Muda
	Kairoshi

218. Which one is not a value of Lean Development?

Agile Certification Study Guide – Answers

Correct	Choice
	Pursue perfection
X	Ensure collective code ownership
	After a project flows, keep improving it
	Balance long-term improvement and short-term improvement

Additional feedback: Collective code ownership is an XP principle.

219. Pick 5 activities that are the responsibilities of the development team in Scrum.

Correct	Choice
X	Provides estimates
	Prioritizes the backlog
	Creates User Stories
X	Commits to the Sprint
	Performs user acceptance
	Facilitates meetings
	Champions Scrum
X	Volunteers for tasks
X	Makes technical decisions
X	Designs software

220. Pick which 4 activities are the responsibilities of the Product Owner in Scrum.

Correct	Choice
	Provides Estimates
X	Prioritize the backlog
X	Create User Stories
	Commit to the Sprint
X	Perform user acceptance
	Facilitate meetings
	Champion Scrum
X	Perform release planning
	Design software

221. Pick which 3 activities are the responsibilities of the Scrum Master in Scrum.

Agile Certification Study Guide – Answers

Correct	Choice
	Provide Estimates
	Prioritize the backlog
	Commit to the Sprint
	Perform user acceptance
X	Facilitate meetings
X	Champion Scrum
	Volunteer for tasks
	Make technical decisions
X	Remove impediments

222. What of the following is not a step in the Value Stream Mapping process?

Correct	Choice
	Define the current state
	Collect data
X	Amplify Learning
	Depict the future state
	Develop an implementation plan

Additional feedback: Amplify Learning is a Lean development principle

223. At completion of iteration planning, the team has finished identifying the tasks they will commit to for the next iteration. Which of the following tools best provides transparency into the progress throughout the iteration?

Correct	Choice
X	Burndown chart
	Gantt chart
	Hours expended chart
	Management baseline chart

Additional feedback: A Sprint Burndown chart will show work completed and working remaining throughout the iteration.

224. A common reason a story may not be estimable is the:

Correct	Choice
X	Team lacks domain knowledge.

Agile Certification Study Guide – Answers

	The story did not include a role.
	Developers do not understand the tasks related to the story.
	Team has no experience in estimating.

Additional feedback: There are three common reasons why a story may not be estimable: 1) Developers lack domain knowledge 2) Developers lack technical knowledge 3) The story is too big.

225. The purpose of a Sprint Retrospective is for the Scrum Team to:

Correct	Choice
	Review stories planned for the next Sprint and provide estimates.
	Demonstrate completed User Stories to the Product Owner.
X	Determine what to stop doing, start doing, and continue doing.
	Individually provide status updates on the User Stories in progress.

226. Question: Which of the following BEST describes ROTI?

Correct	Choice
	Measure of product backlog items (PBI) remaining
	Measure of quality of features delivered in an iteration
	Measure of required effort to complete an iteration
X	Measure of the effectiveness of the retrospective meeting

Additional feedback: Return on Time Invested (ROTI) is used to measure the effectiveness of the retrospective meetings from the team members' perspective.

227. What Agile concept expresses delivering value in slices rather than in layers/stages?

Correct	Choice
	Definition of Done
	Value Mapping
X	Sashimi
	Lean Value

228. In the Kano Model of customer satisfaction, this type of feature makes a product unique from its competitors and contributes 100% to positive customer satisfaction:

Correct	Choice
X	Excitement

Agile Certification Study Guide –Answers

Performance
Must-have
Threshold

Additional feedback: Excitement attributes are for the most part unforeseen by the customer but may yield the most satisfaction.

229. Which of the following is NOT a principle from the Agile Manifesto?

Correct	Choice
	Our highest priority is to satisfy the customer through early and continuous delivery of valuable software.
	Business people and developers must work together daily throughout the project.
X	Continuous creation of technical debt and good design enhances agility.
	Working software is the primary measure of progress.

Additional feedback: Continuous creation of technical debt and good design enhances agility is NOT a principle of the Agile Manifesto. The real one is, "continuous attention to technical excellence and good design enhances agility."

230. Which chart shows the total number of story points completed through the end of each iteration?

Correct	Choice
	Iteration Burndown chart
X	Cumulative story point Burndown chart
	Daily Burndown chart
	Burnup chart

Additional feedback: A cumulative story point burndown chart shows the total number of story points completed through the end of each iteration.

231. What is the order the hierarchy of product definition?

Correct Order
Product Vision
Product Roadmap
Theme
Epic
User Story

Agile Certification Study Guide –Answers

Task

232. What Agile planning artifact is updated minimally once a year by the Product Owner?

Correct	Choice
X	Product Vision
	Product Roadmap
	Release Plan
	Sprint Plan
	Daily Plan

Additional feedback: The Product Vision is updated annually by the Product Owner and stakeholders.

233. What Agile planning artifact should be updated at minimum semi-annually?

Correct	Choice
	Product Vision
X	Product Roadmap
	Release Plan
	Sprint Plan
	Daily Plan

Additional feedback: The Product Roadmap should be updated at least twice a year by the Product Owner.

234. What Agile planning artifact is created by the Product Owner and the development team?

Correct	Choice
	Product Vision
	Product Roadmap
	Release Plan
X	Sprint Plan
	Daily Plan

Additional feedback: The Sprint Plan is created by the Product Owner and the Development Team.

235. DSDM uses MoSCoW technique to create the prioritized requirements list. In MoSCoW technique, 'M' stands for:

Agile Certification Study Guide – Answers

Correct	Choice
	Most useful
X	Must have
	Must not have
	Minimum marketable feature

Additional feedback: MoSCoW is an acronym for: Must Have, Should Have, Could Have, Won't Have

236. Based upon this Burndown chart, is this project ahead of schedule or behind schedule?

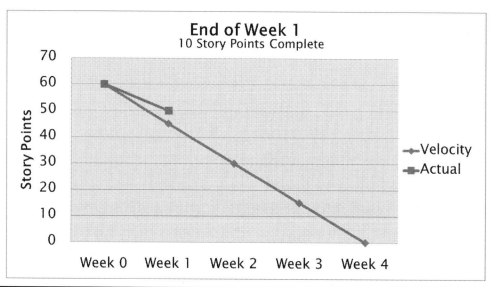

Correct	Choice
	Ahead of schedule
X	Behind schedule

Additional feedback: If the actual line is above the velocity line, the Burndown chart shows that the project is behind schedule.

237. Which of the following occurs in the first Sprint?

Correct	Choice
	Create a project plan
X	Develop a shippable piece of functionality
	Complete your reference architecture
	Develop the Product Roadmap

238. With multiple Scrum teams, you should have a separate product backlog.

Agile Certification Study Guide – Answers

Correct	Choice
	True
X	False

Additional feedback: There is only one product backlog when multiple Scrum teams are working on the same product.

239. What is the best definition of "Done"?

Correct	Choice
	Whatever will please the Product Owner
	It is determined by the Scrum Master
	The product has passed QA and has all of the required release documentation
X	The definition of "Done" is one that would allow the development work to be ready for a release

Additional feedback: The definition of "Done" is one that would allow development work to be ready for a release.

240. Which of the following does NOT describe Scrum?

Correct	Choice
	Simple to understand
	A lightweight framework
	Difficult to master
X	A process or a technique for building products

Additional feedback: Scrum is not a process or a technique for building products; rather, it is a framework within which you can employ various processes and techniques.

241. Scrum is NOT:

Correct	Choice
	A set of software project management principles
	Founded on empirical process control theory, or empiricism.
	A process framework used to manage the development of products
X	Designed for static requirements

Additional feedback: Scrum is intended for use with adaptive requirements not static requirements.

242. Which of the following is NOT part of the Scrum Framework?

Agile Certification Study Guide –Answers

Correct	Choice
	Roles
	Events
X	Characteristics
	Artifacts

Additional feedback: The Scrum framework consists of Scrum Teams and their associated roles, events, artifacts, and rules.

243. What are the three pillars of Scrum?

Correct	Choice
X	Transparency, Inspection, and Adaptation
	Transparency, Inspection, and Empiricism
	Transparency, Acceptance and Adaptation
	Retrospectives, Inspection, and Adaptation

244. What does Scrum mean by Transparency?

Correct	Choice
	Users can perform code reviews at any time
	Documentation is available to anyone
	All team members sit in a visible location
X	The process is understandable by all stakeholders

Additional feedback: Transparency requires work be performed, documented, or presented in a common standard so observers can share a common understanding of all document artifacts and the progress of work.

245. When does Adaptation occur in Scrum?

Correct	Choice
	At the Sprint Review
	During Sprint Planning
	In the daily Scrum
X	At all four formal Scrum events
	As Part of the Sprint Retrospective

246. Who is NOT part of the Scrum Team?

Agile Certification Study Guide – Answers

Correct	Choice
	Product Owner
	Scrum Master
X	Customer
	Development Team

Additional feedback: The Scrum Team consists of a Product Owner, the Development Team, and a Scrum Master. The customer is not part of the Team.

247. A cross-functional team in Scrum consists of which types of team members?

Correct	Choice
	A specialist in QA
	An architect
	A release manager
X	Anyone with the skills to accomplish the work

Additional feedback: Cross-functional teams have all competencies needed to accomplish the work without depending on others not part of the team.

248. Scrum is both an iterative and incremental Agile process.

Correct	Choice
X	True
	False

Additional feedback: Scrum is both an iterative and incremental Agile process.

249. When does Inspection occur?

Correct	Choice
	Throughout the Sprint
	Only at the end of the Sprint
	Whenever the Product Owner wishes
X	Frequently, but not so often that it gets in the way of work

Additional feedback: Inspection occurs frequently, but only enough to detect variances in the work.

250. The Product Owner is the sole person responsible for managing the Product Backlog.

Agile Certification Study Guide – Answers

Correct	Choice
X	True
	False

251. Who is responsible for maximizing the value of the product?

Correct	Choice
	Senior Executives
X	The Product Owner
	The Scrum Master
	The Development Team

252. The Product Owner does not have to be a single person but may be a committee or a shared responsibility between multiple individuals.

Correct	Choice
	True
X	False

Additional feedback: The Product Owner is one person, not a committee. The Product Owner may represent the desires of a committee in the Product Backlog, but those wanting to change a Product Backlog item's priority must address the Product Owner.

253. No one, not even the Scrum Master, tells the development team how to build the product.

Correct	Choice
X	True
	False

Additional feedback: They are self-organizing. No one (not even the Scrum Master) tells the Development Team how to turn Product Backlog into Increments of potentially releasable functionality.

254. The development team should have a lead developer to ensure the work is properly executed.

Correct	Choice
	True
X	False

Additional feedback: The Scrum Team is cross-functional and self-organizing. There is no lead developer.

Agile Certification Study Guide – Answers

255. The optimum size of the Scrum Team is:

Correct	Choice
	7
X	Between 3 and 9
	5
	It depends

Additional feedback: Less than 3 may not be an efficient team and greater than 9 makes coordination difficult. Any number in within the range of 3 - 9 is acceptable.

256. The Product Owner and Scrum Master are never part of the Development Team.

Correct	Choice
	True
X	False

Additional feedback: The Scrum Master and Product Owner can be part of the Development Team if they help create the product.

257. The Scrum Master as a Servant Leader is in service to which of the following?

Correct	Choice
	The Development Team
	The Organization
	The Product Owner and the Development Team
X	The Organization, the Product Owner and the Development Team

258. Which one of the following is NOT a Scrum Event?

Correct	Choice
	Sprint
	Daily Scrum
	Sprint Review
X	Weekly Status
	They are all Scrum events.

259. A new Sprint starts immediately following the previous Sprint.

Correct	Choice
X	True

Agile Certification Study Guide – Answers

False

260. Sprints lengths can vary each Sprint as long as they don't exceed a month.

Correct	Choice
	True
X	False

Additional feedback: Sprint lengths should remain a consistent length throughout a project.

261. Put the following in order of first occurrence in a Sprint.

Correct Order
Sprint Planning
Daily Scrum
Sprint Review
Sprint Retrospective

262. All of the following are true about change during a Sprint EXCEPT:

Correct	Choice
X	Changes can be made that impact the Sprint goal
	The development team can change tasks in the Sprint backlog
	The Product Owner is the only person that can add or remove a User Story in the Sprint Backlog
	Change may occur as scope is clarified between the Product Owner and the Development Team

Additional feedback: Changes that impact the Sprint Goal are not allowed. The other listed changes are allowed.

263. The Scrum Team and Development Team are the same thing.

Correct	Choice
	True
X	False

Additional feedback: The Scrum Team includes the Development Team, Scrum Master and Product Owner.

264. In Scrum, Sprints are never longer than a calendar month.

Agile Certification Study Guide – Answers

Correct	Choice
X	True
	False

Additional feedback: Sprints are limited to one calendar month. When a Sprint's horizon is too long the definition of what is being built may change, complexity may rise, and risk may increase.

265. Who can cancel a Sprint?

Correct	Choice
	The Development Team
	Executive Stakeholders
X	The Product Owner
	The Scrum Master

Additional feedback: Only the Product Owner has the authority to cancel a Sprint.

266. What happens if the customer no longer wants the feature that the Sprint Goal intended to meet?

Correct	Choice
	The Development Team should determine if there is value in the Sprint
	The Executive Stakeholders should determine if the Sprint should continue
X	The Product Owner should cancel the Sprint
	The Scrum Master should cancel the Sprint

Additional feedback: If the Sprint goal is no longer valid, the Product Owner may cancel a Sprint if the Sprint backlog is no longer valuable.

267. How long is the Sprint Planning meeting?

Correct	Choice
	4 hours
	8 hours
X	Depends on the length of the Sprint
	3 hours

Additional feedback: The duration of the Sprint planning meeting varies based upon Sprint length. For a one month Sprint, the planning meeting is 8 hours, two hours per week of Sprint.

Agile Certification Study Guide – Answers

268. In the first part of the Sprint Planning meeting, what is NOT accomplished?

Correct	Choice
	Items are selected from the Product Backlog
	The Development Team decides how much work can be accomplished
	The Scrum Team defines the Sprint Goal
X	The Tasks are defined

Additional feedback: The Tasks are defined in the second part of the Sprint planning meeting.

269. No one but the Scrum Team attends the Sprint Planning meeting.

Correct	Choice
	True
X	False

Additional feedback: The Development Team may also invite other people to attend in order to provide technical or domain advice.

270. Which three of the following points about the daily Scrum are TRUE?

Correct	Choice
X	It is time-boxed
X	It is held at the same place and time every day
	The Product Owner provides an update
X	The Scrum Master enforces the rule that only Development Team members participate

Additional feedback: All are correct except that the Product Owner does not speak at the daily Scrum unless they have a task on a User Story

271. Which of the following are TRUE about the Sprint Review? (Choose two).

Correct	Choice
	It should be a formal presentation
X	Stakeholders may attend
X	The Product Owner presents what backlog items are "Done"
	The Scrum Master demonstrates the product

Additional feedback: The Sprint Review is an informal event where the Development Team demonstrates the product. Other stakeholders may attend, and the Product Owner presents the backlog items that are Done.

Agile Certification Study Guide – Answers

272. The feedback from the Sprint Review impacts the next Sprint planning meeting.

Correct	Choice
X	True
	False

273. Which of the following are TRUE about the Sprint Retrospective? (Choose 2)

Correct	Choice
X	It is three hours for a one month Sprint
	It occurs before the Sprint Review
X	It is an opportunity to inspect the people, relationships, process, and tools in the last Sprint
	It is the only time improvements are made during a Sprint

Additional feedback: It occurs after the Sprint Review. This meeting is only one of the opportunities to Inspect and Adapt. The duration is three hours long for a one month Sprint.

274. Match the following:

Correct	Choice
Scrum Event	Daily Scrum
Scrum Artifact	Product Backlog
Scrum Role	Product Owner
Not Scrum	Gantt Chart

275. The Product Backlog is baselined at the start of the project and not changed for at least three Sprints.

Correct	Choice
	True
X	False

Additional feedback: The Product Backlog is reviewed during each Sprint and updated if necessary.

276. Who is responsible for ordering the Product Backlog?

Correct	Choice
	Senior Executives
X	The Product Owner

	The Scrum Master
	The Development Team

277. Which item is NOT an attribute of the Product Backlog?

Correct	Choice
	Description
	Order
	Estimate
	Value
X	Owner

Additional feedback: Owner is not an attribute of the Product Backlog; all the others items are.

278. Backlog Grooming and Backlog Refinement are the same thing.

Correct	Choice
X	True
	False

279. Which of the following 2 statements are TRUE about Product Refinement?

Correct	Choice
X	Should take no more than 10% of the Development Team's time
	The Scrum Master facilitates these sessions
X	Multiple Scrum Teams may participate in this process
	The Product Owner is responsible for all estimates

Additional feedback: Product refinement should take no more than 10% of the Development Team's time. The Product Owner and Development Team collaborate to refine the backlog. It should include all Scrum Teams working on a product. The Development Team is responsible for estimating.

280. Who tracks work remaining in the Product Backlog?

Correct	Choice
	The Development Team
	The Scrum Master
X	The Product Owner
	Senior Executives

Agile Certification Study Guide – Answers

281. Who can change the Sprint Backlog during a Sprint?

Correct	Choice
	Senior Executives
	The Product Owner
	The Scrum Master
X	The Development Team

Additional feedback: Only the Development Team can change the Sprint Backlog during a Sprint. The Product Owner is responsible for the Product Backlog.

282. If there are multiple Scrum teams working on a product, each needs its own definition of Done.

Correct	Choice
	True
X	False

Additional feedback: If there are multiple Scrum Teams working on the same product, they each need the same definition of "done."

283. To truly adopt Scrum, you must pick and choose what roles, artifacts, events, and rules are right for your organization.

Correct	Choice
	True
X	False

Additional feedback: Scrum's roles, artifacts, events, and rules are immutable and although implementing only parts of Scrum is possible, the result is not Scrum. Scrum exists only in its entirety and functions well as a container for other techniques, methodologies, and practices.

284. When was Scrum first introduced?

Correct	Choice
	In 2001 with the Agile Manifesto
X	In 1995 at a conference presentation
	At General Electric as part of a Lean approach
	In 2000 along with Extreme Programming

285. Scrum is a container for other techniques and methodologies.

Agile Certification Study Guide – Answers

Correct	Choice
X	True
	False

Additional feedback: Scrum is a container or framework for other product development approaches, e.g. XP, TDD, etc.

286. What type of process control is Scrum?

Correct	Choice
	Classical
X	Empirical
	Inspection
	Adaptive

Additional feedback: Scrum is an empirical process since it provides control through frequent inspection and adaptation for processes.

287. The Scrum Master is a management role?

Correct	Choice
X	True
	False

Additional feedback: The Scrum Master has to be a member of management to be able to remove impediments for the Development Team.

288. There must be a release every Sprint.

Correct	Choice
	True
X	False

Additional feedback: Each Sprint should create a shippable piece of functionality, but is up to the Product Owner when to release.

289. How often should Development Team members change?

Correct	Choice
	No more than every three Sprints
	Never
	Each Sprint
X	As needed

Agile Certification Study Guide – Answers

Additional feedback: Teams members come and go for a variety of reasons outside of anyone's control. Each time someone leaves there will be a short-term reduction in productivity.

290. What is a time-boxed event?

Correct	Choice
	It happens at the same time as a conflicting task
X	It has a maximum duration
	It has a minimum duration
	It has a fixed place and time

291. When is a Sprint finished?

Correct	Choice
	When the definition of "Done" is met
	When the Product Owner accepts the increment
X	When the time-boxed duration is met
	When the work remaining is zero

Additional feedback: The only time a Sprint is finished is when the time-boxed duration is met. If the team completes the Sprint backlog, the Product Owner adds more work to the Sprint.

292. Who updates work remaining during the Sprint?

Correct	Choice
	Senior Executives
	The Product Owner
	The Scrum Master
X	The Development Team

Additional feedback: The Development Team tracks this total work remaining at least for every Daily Scrum to project the likelihood of achieving the Sprint Goal.

293. Identify all members of a Scrum Team:

Correct	Choice
	Customer
	Stakeholder

Correct	Choice
X	Product Owner
X	Scrum Master
	Project Manager
X	Development Team

Additional feedback: The Scrum Team consists of the Scrum Master, Product Owner and the Development Team.

294. Who is responsible for the Project Plan and Gantt Chart in Scrum?

Correct	Choice
	Project Manager
	Scrum Master
	Product Owner
X	No Scrum role

Additional feedback: There is neither a project manager nor Gantt Charts in Scrum.

295. How long is a Sprint Review?

Correct	Choice
	2 hours
	4 hours
X	It depends on the length of the Sprint

Additional feedback: A Sprint Review is 4 hours for a full month Sprint, but less if the Sprint is shorter.

296. If the Sprint Backlog cannot be completed in a Sprint, who resolves the issue?

Correct	Choice
	Product Owner
	Scrum Master
	Development Team
X	Both the Product Owner and Development Team

Additional feedback: The Development Team and the Product Owner collaborate to determine the impact on the Sprint Backlog.

297. If a customer really wants a feature added to a Sprint, how should the Development Team respond?

Agile Certification Study Guide – Answers

Correct	Choice
	Add the feature into the current Sprint backlog
	Escalate to the Scrum Master
	Add the item to the Product Backlog for prioritization in the next Sprint
X	Ask the Product Owner to work with the customer

Additional feedback: The Development Team should ask the Product Owner to work with the customer to resolve the issue.

298. When does a Sprint get canceled or end early?

Correct	Choice
	When the Sprint backlog is complete
X	When the Sprint Goal cannot be met
	When the definition of "Done" is met
	When a key resource is out sick

299. How long is the time-box for the daily Scrum?

Correct	Choice
	It depends
	5 minutes per person on the Development Team
X	15 minutes
	Whatever the Team decides

300. How does the Scrum Master provide the most value to the Team?

Correct	Choice
	By facilitating discussions between the Product Owner and the Development Team
	Ensuring time-boxes are kept
X	Removing impediments to the Development Team
	Scheduling Scrum events

301. Select the statements that are TRUE about the Product Owner. (Choose two)

Correct	Choice
X	The Product Owner can clarify the backlog during the Sprint
	The Product Owner estimates the size of the Sprint backlog

Agile Certification Study Guide – Answers

Correct	Choice
X	The Product Owner prioritizes the Product backlog
	The Product Owner defines the Sprint Goal before the Sprint Planning meeting

Additional feedback: The Product Owner is responsible for the ordering of the product backlog and can provide clarification during the Sprint. He or she does not estimate the work, nor should they have the Sprint goal defined before the Sprint Planning meeting.

302. Who creates the Sprint Goal?

Correct	Choice
	The Development Team
	The Scrum Master
	The Product Owner
X	The entire Scrum Team

Additional feedback: The Scrum Team consisting of the Development Team, Scrum Master, and Product Owner creates the Sprint Goal.

303. In Scrum, the development team decides which events or ceremonies take place during a Sprint.

Correct	Choice
	True
X	False

Additional feedback: In Scrum, ALL events take place. "Failure to include any of these events results in reduced transparency and is a lost opportunity to inspect and adapt."

304. The Scrum Master is a participant in the Sprint Retrospective.

Correct	Choice
X	True
	False

Additional feedback: The Scrum Master is a participant in the Sprint Retrospective.

305. If the Development Team does not have all the skills to accomplish the Sprint Goal, the Scrum Master should:

Correct	Choice
	Cancel the Sprint
	Stop using Scrum
X	Have the development team determine the definition of "Done" and

	work through the Sprint backlog
	Remove the impacted stories from the Sprint backlog

Additional feedback: The Development Team needs to self-organize to determine what can be accomplished with the team members it has; they develop the additional skills over time to improve the definition of Done.

306. The Project Manager plays the following role in Scrum:

Correct	Choice
	Collects the status from the Scrum Master
	Updates the Burndown chart
	Creates the release plan
X	There is no project manager role in Scrum

307. What is the purpose of the Sprint Review? (Choose three)

Correct	Choice
X	To collaborate with stakeholders
X	To inspect and adapt
	To provide status on the Sprint
X	To demonstrate what is "Done"

Additional feedback: Inspect and Adapt, collaborate with stakeholders, and demonstrate "Done" are performed during the Sprint Review.

308. Scrum dictates the use of User Stories.

Correct	Choice
	True
X	False

Additional feedback: User Stories are an Agile requirement gathering technique, but not an artifact of the Scrum framework.

309. Scrum is a software development methodology.

Correct	Choice
	True
X	False

Additional feedback: Scrum is a process control methodology and is not used only for software development projects.

310. If the Development Team doesn't like the time of the daily Scrum, what should the Scrum Master do?

Correct	Choice
	Find a time that is open on everyone's calendar
X	Let the Development Team come up with a new time
	Ask the Team to try the existing time for one Sprint
	Tell them that Scrum is immutable and that they need to stick to it

Additional feedback: Let the Development Team come up with a new time since they are self-organized.

311. The backlog is ordered by:

Correct	Choice
X	The needs of the Product Owner
	Risk
	Complexity
	Size

312. Who is responsible for maximizing the value of the product backlog?

Correct	Choice
	The Customer
	The Scrum Master and Product Owner
X	The Product Owner
	The Development Team and Product Owner

313. What happens if all the necessary testing doesn't occur in a Sprint?

Correct	Choice
	The User Story is moved to the next Sprint
	Additional testers are added in the next Sprint
X	A risk of not creating a potentially shippable product occurs
	The Burndown chart is updated

314. Pick roles that support the Scrum Master in removing impediments. (Choose two.)

Correct	Choice
X	The Development Team

Agile Certification Study Guide – Answers

X	Senior Management
	The Product Owner
	The Customer

315. Match each of the following items with its associated time-boxed duration for a one-month Sprint.

Correct	Choice
Sprint Review	4 hours
Sprint Retrospective	3 hours
Sprint Planning	8 hours
Sprint	1 month
Daily Scrum	15 minutes

316. Match the activity (on the right) to the Scrum event (on the left).

Correct	Choice
Sprint Planning	Sprint Goal creation
Sprint Retrospective	Adapt the definition of "Done"
Daily Scrum	Inspect and adapt
Sprint Review	Demonstrate Functionality

317. What Scrum event or artifact supports daily inspection and adaptation?

Correct	Choice
	Product Backlog
	Sprint Backlog
	Sprint
X	Scrum
	Working Product Increment

Additional feedback: The Development Team uses the daily Scrum to inspect progress toward the Sprint Goal and to inspect how progress is trending toward completing the work in the Sprint Backlog

318. What Scrum event or artifact is the single source of requirements for any changes to be made to the product?

Correct	Choice
X	Product Backlog

	Sprint Backlog
	Sprint
	Scrum
	Working Product Increment

Additional feedback: The Product Backlog is an ordered list of everything that might be needed in the product and is the single source of requirements for any changes to be made to the product.

319. What Scrum event or artifact is the set of items selected for the Sprint, plus a plan for delivering the product Increment and realizing the Sprint Goal?

Correct	Choice
	Product Backlog
X	Sprint Backlog
	Sprint
	Scrum
	Working Product Increment

Additional feedback: The Sprint Backlog is the set of Product Backlog items selected for the Sprint, plus a plan for delivering the product Increment and realizing the Sprint Goal.

320. The Scrum Master's job is to work with the Scrum Team and the organization to increase the awareness of the artifacts. Which pillar of Scrum does this represent?

Correct	Choice
X	Transparency
	Inspection
	Adaptation

Additional feedback: The Scrum Master must work with the Product Owner, Development Team, and other involved parties to understand if the artifacts are completely transparent.

321. Scrum users must frequently review Scrum artifacts and progress toward a Sprint Goal to detect undesirable variances. Which pillar of Scrum does this represent?

Correct	Choice
	Transparency
X	Inspection
	Adaptation

322. Based upon this Burndown chart, is this project ahead of schedule or behind schedule?

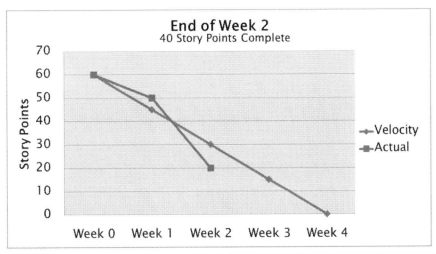

Correct	Choice
X	Ahead of schedule
	Behind schedule

Additional feedback: If the actual line is below the velocity line, the Burndown chart shows that the project is ahead schedule.

323. Pick the two PMLC models that are based upon the Agile Project Management (APM) approach:

Correct	Choice
	Linear
X	Adaptive
	Incremental
X	Iterative

324. Which of the following is NOT a characteristic of an Adaptive PMLC Model?

Correct	Choice
	Iterative Structure
X	Clear up front requirements
	Mission Critical Projects
	JIT Planning

Additional feedback: Adaptive models have minimal information at the beginning of the project.

Agile Certification Study Guide – Answers

325. This management approach is based on knowing well defined goals but not the means for a solution.

Correct	Choice
	Traditional Project Management
	Emertxe Project Management
	Extreme Project Management
X	Agile Project Management

326. This Emertxe Project Management (MPx) approach is when neither a goal nor solution is clearly defined.

Correct	Choice
	True
X	False

Additional feedback: The Emertxe Project Management approach is when the solution is well defined, however, the goal is not defined.

327. Every Project Management Life Cycle (PMLC) has a sequence of processes that include these phases:

    Scoping
    Planning
    Launching
    Monitoring & Controlling
    Closing

Correct	Choice
X	True
	False

Additional feedback: Every valid project management life cycle must include each of these processes one or more times.

328. Which of the following is a weakness of an Adaptive PMLC Model?

Correct	Choice
	Does not waste time on non-value-added work
	Does not waste time planning uncertainty
X	Cannot identify what will be delivered at the end of the project
	Avoids all management issues processing scope change requests

329. Sequence the core practices of Kanban in order of execution.

Agile Certification Study Guide – Answers

Correct Order
Visualize the workflow
Limit WIP
Manage the flow
Make process policies explicit
Implement Feedback Loops
Improve collaboratively

330. The number of days needed between feature specification and production delivery is called:

Correct	Choice
X	Cycle time
	Calendar time
	Ideal time
	Real time

331. The number of days needed between customer request and production delivery is called:

Correct	Choice
	Cycle time
X	Lead time
	Ideal time
	Real time

332. Classes of Services in Kanban are used to:

Correct	Choice
	Support estimation for Kanban Cards
X	Prioritize the queue by risk
	All of the above
	Ensure WIP limits are realistic

Additional feedback: Classifying classes of service are typically priorities defined based on business impact and cost of delay.

333. The purpose of Work in Progress (WIP) limits is to prevent the unintentional accumulation of work, so there isn't a bottleneck.

Agile Certification Study Guide – Answers

Correct	Choice
	False
X	True

Additional feedback: One of the core properties in the Kanban method is that Work in Progress is limited. Limiting WIP is to match team's development capacity and to prevent bottlenecks.

334. The following is a picture of which of the following Information Radiators?

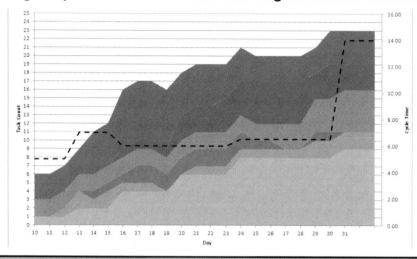

Correct	Choice
	Burndown Chart
	Kanban Tracking System
X	Cumulative Flow Diagram
	Burnup Chart

335. The measure of productivity of a Kanban team is:

Correct	Choice
	Cycle time
	Lead Time
	Work in Progress
	Velocity
X	Throughput

336. Kanban cards should always be written using User Stories.

Correct	Choice
	True

Agile Certification Study Guide – Answers

X	False

Additional feedback: There are many formats for Kanban Cards. Using User Stories is just one approach.

337. A term used to describe the work that can be delivered which meets the business requirements without exceeding them. (Choose Two)

Correct	Choice
	Epic
X	Minimum Viable Product
	Theme
	User Story
X	Minimum Marketable Features

338. Order the 5 focusing steps of the Theory of Constraints.

Correct Order
Identify the System Constraint
Decide How to Exploit the Constraint
Subordinate Everything Else
Elevate the Constraint
Go Back to Step 1, Repeat

339. This role champions the products, provides the budget and supports the Scrum Master in removing impediments

Correct	Choice
	Subject Matter Expert
	Product Owner
X	Business Owner
	Project Manager

340. The behavior where the Scrum Team focuses on one or more stories until they are done is called:

Correct	Choice
	Collaboration
X	Swarming
	Pair-programming

Agile Certification Study Guide – Answers

	Sprinting

341. Match the following roles on the right to the RASCI on the left:

Correct	Choice
Accountable	Product Owner
Consulted	Subject Matter Experts
Informed	Stakeholders and Business Owner
Supportive	Scrum Master
Responsible	Team Members

342. The product owner should spend at least 3 hours per day with the development team?

Correct	Choice
X	True
	False

343. Which role is external to the Scrum Team but provides a skills the does not exist on the Team?

Correct	Choice
X	Subject Matter Expert
	Team Member
	Scrum Master
	Project Manager

344. Pick the 3 common Kanban Katas.

Correct	Choice
X	Daily Standup Meeting
	Iteration Demo
X	Improvement Kata
	Sprint Retrospective
	Weekly Status
X	Operations Review

345. Which of the following is NOT considered an enterprise Agile method?

Agile Certification Study Guide – Answers

Correct	Choice
	DAD
X	XP
	LeSS
	SAFe

346. Which Agile enterprise framework adopts and tailors methods such as Scrum, Extreme Programming (XP), and Agile Modeling (AM) in order to support scaling.

Correct	Choice
X	Disciplined Agile Delivery (DAD)
	Agile Delivery Framework (ADF)
	DSDM
	Crystal

Additional feedback: Disciplined Agile Delivery DAD adopts and tailors proven strategies from existing methods such as Scrum, Extreme Programming (XP), Agile Modeling (AM), Unified Process (UP), Kanban, Outside-in software development, and Agile Data (AD).

347. This Agile methodology's properties include Focus, Osmotic Communication and Project Safety.

Correct	Choice
	Scrum
X	Crystal Clear
	Extreme Programming
	Kanban

Additional feedback: The Crystal family of methodologies focuses on efficiency and habitability as components of project safety. Crystal Clear focuses on people, not processes or artifacts.

348. Which of the following frameworks has the following practices?

> Supports an envision, explore, adapt culture
> Supports a self-organizing, self-disciplined team
> Promotes reliability and consistency to the extent possible given the level of project uncertainty
> Provides management checkpoints for review

Correct	Choice
	Dynamic Systems Development Method (DSDM)

Agile Certification Study Guide –Answers

	Crystal Clear
	Feature Driven Development (FDD)
X	Agile Delivery Framework

349. Which of the following is NOT a prioritization technique?

Correct	Choice
	User Story Mapping
	Kano analysis
	Minimally Marketable Features (MMF)
	Kitchen Prioritization
X	Planning Poker

Additional feedback: Planning Poker is an estimation technique.

350. Sequence the steps to User Story Mapping.

Correct Order
Product Roadmap Definition
User Story Definition
User Story Decomposition
Release Planning

351. Choose which three statements are true about Approved Iterations:

Correct	Choice
X	Meets the definition of Done
	The Architect has approved it
X	It is communicated to all Team members and stakeholders
X	As a result, the Product Owner updates Roadmaps and Release Plans
	There is no technical debt

352. "Fail Sooner" is a benefit of Incremental Development.

Correct	Choice
X	True
	False

Agile Certification Study Guide – Answers

Additional feedback: Incremental Development mitigates the risk of project failure by not meeting the stakeholder's requirements. Deliver in chunks and expect feedback: "Fail Sooner"

353. In what order do you select requirements to work on in a risk adjusted backlog?

Correct Order
Low Risk High Value
High Risk High Value
Low Risk Low Value
High Risk Low Value

354. Non-functional requirements should be written as user stories whenever possible.

Correct	Choice
	True
X	False

Additional feedback: NFRs don't have to be written as User Stories. Instead they are often part of the Definition of Done or part of the Acceptance Criteria

355. CRUD is an acronym that defines a way to split stories. What does CRUD stand for?

Correct	Choice
X	Create/Update/Delete
	Capture/Replace/Update/Define
	Create/Review/Update/Done
	Create/Replace/Update/Define

Additional feedback: Create/Update/Delete

356. Match the story types on the left to their definition on the right:

Correct	Choice
Architecturally Significant	Stories are functional Stories that cause the Team to make architectural decisions
Analysis	Stories that are created to find other stories
Infrastructure	Stories to improve the infrastructure the team is using
Spike	Stories that figure out answers to tough technical or design problems

357. Operations and Maintenance staff should not be part of the Agile team.

Agile Certification Study Guide – Answers

Correct	Choice
	True
X	False

Additional feedback: The approach of DevOps suggests that Operations and Maintenance staff should be incorporated into Agile Teams.

358. Which of the following is NOT a metric that measures the performance of DevOps in Agile?

Correct	Choice
	Release date adherence percentage
	Defects attributable to platform/support requirements
	Percentage of NFRs met
X	Business value realized per release
	Percentage increase in the number of releases

Additional feedback: Business value metrics are attributed to the business and are captured by the product owner and business owner, not DevOps.

359. Which of the following is NOT a KPI (key performance indicator) of Agile?

Correct	Choice
	Actual Stories Completed vs. Committed Stories
	Quality Delivered to Customers
	Team Enthusiasm
X	Team Attendance
	Team Velocity
	Technical Debt Management

Additional feedback: Team Attendance is not a KPI of Agile. Agile teams are self-managing so less focus is spent on micro-managing time in the office, etc.

360. Your project management office (PMO) has suggested your project could benefit from some self-assessment work at the next retrospective. Which of the following benefits would they most likely be looking to achieve from a self-assessment?

Correct	Choice
	Assess compatibilities for pair programming assignments
X	Improve personal and team practices
	Identify personal traits for human resources counseling

Agile Certification Study Guide – Answers

| | Gain insights for salary performance reviews |

Additional feedback: The benefits would they most likely be looking to achieve from a self-assessment would be to improve personal and team practices.

361. The definition of an Epic can be written using the acronym CURB. What does CURB stand for.

Correct	Choice
	Create/Update/Replace/Big
	Compound/Unknown/Risky/Basic
	Complicated/Unusual/Really Big
X	Complex/Unknown/Risky/Big

Additional feedback: Complex/Unknown/Risky/Big

362. The Japanese terms for an Agile developmental mastery model are:

Correct	Choice
	Kaizen
	Muda
	Sashimi
X	Shu-Ha-Ri

Additional feedback: Su-Ha-Ri is an Agile development mastery model taken from Japanese martial arts.

363. Which development mastery model of skill acquisition lies in helping the teacher understand how to assist the learner in advancing to the next level?

Correct	Choice
	Shu-Ha-Ri
	Kaizen
	Tuckman
X	Dreyfus

364. Which of the following is not part of Agile Discovery?

Correct	Choice
	Document business outcomes that are quantifiable and measurable
	Outline a plan for the technical and business architecture/design of the solution

Agile Certification Study Guide – Answers

	Describe essential governance and organization aspects of the project and how the project will be managed
X	Define the tasks that the team will perform during an iteration

Additional feedback: Agile Discovery occurs before the project begins. Task planning occurs with the team during iteration planning.

365. Agile Analysis is a phase in the lifecycle of an Agile project.

Correct	Choice
X	False
	True

Additional feedback: Agile Analysis isn't a phase in the lifecycle of your project. It is an ongoing and iterative activity.

366. Pick the THREE statements that are true about Agile Analysis.

Correct	Choice
X	It is a highly evolutionary and collaborative process
	It occurs at the beginning and end of a project
	It only includes the project team
X	It is communication rich
X	It explores the problem statement

367. It is not possible to have a fixed price contract in Agile.

Correct	Choice
	True
X	False

Additional feedback: Though a time and materials contract is an easier alternative for an Agile contract, the reality is that fixed-price contracts are often necessary. In a perfect world, they should be limited to Agile discovery. If you are required to have fixed price contracts, create separate contracts for each iteration.

368. Match the traditional contract model of the left with the Agile alternative on the right.

Correct	Choice
Analysis, design, development and testing occur sequentially.	There is concurrent design and development.
No value delivered until the entire project has been completed.	Value delivered at the end of every Sprint.

Agile Certification Study Guide – Answers

There is no attempt to control the order in which the requirements are tackled.	The highest risk and highest value items are tackled first.
Success is measured by reference to conformance with the plans.	Success is measured by reference to the realization of the desired business outcomes.
Changes 'controlled' by means of the change control mechanism.	Change is accommodated within the non-contractual product backlog.

369. What is the Japanese business philosophy focused on making constant improvements?

Correct	Choice
	Shu-Ha-Ri
	Sashimi
	Aikido
X	Kaizen

370. Which of the following is an Agile improvement technique to address issues continuously, e.g. after daily stand-up?

Correct	Choice
	Retrospectives
	Verification Sessions
X	Intraspectives
	Futurespectives

Additional feedback: Intraspectives: Discussions within the Sprint to address issues. Typically as a result of issues raised during the daily Scrum

371. Similar to inspect and adapt in Scrum, this can be represented as Build, Measure, Learn.

Correct	Choice
	Six Sigma
	Kaizen
	DMAIC
X	Agile Learning Cycle

Additional feedback: Learning cycles assume that we are continually engaging in iterative cycles of learning where doing is connected to observing to reflecting to improving and then repeating. In Agile, this can be represented as Build, Measure, Learn.

Agile Glossary and Web Resources

This content is based upon to the knowledge areas outlined in the most recent (2015) PMI-ACP Examination Content Outline. Although this Guide is not an all-inclusive tool for exam preparation, it addresses many of the key topics covered in the test preparation reference materials suggested by PMI as well as the more common Agile practices. This guide, along taking online practice exams are valuable resources for studying for the Exam. For **an online version** of our PMI-ACP Practice Exam with free samples, go here: http://bit.ly/practice-pmi-acp. For a **free download of this section** of the Study guide, go to: http://bit.ly/ACPStudyGuide. This guide, along taking online practice exams are valuable resources for studying for the Exam. For an online version of our PMI-ACP Practice Exam with free samples, click here: http://bit.ly/practice-pmi-acp . If you have not attended any PMI-ACP training, **consider our online narrated Self-Study training course with over 250 practice questions** embedded in each training module: http://bit.ly/ACPSelfStudy

If you see the [icon] icon beside a term, it is an indicator there was one or more questions related to this term on my version of the exam.

Note: The following Agile terms and definitions are primarily excerpts from web pages or books. As you will see while you study and even when you are taking the Exam, that terms and concepts are defined differently between experts and authors. At times, they even contradict each other. I attempted to capture the content which I believe represents the most common understanding of an Agile term or concept. Lastly, I do not retain the copyright for any of this work. All copyrights reside with their respective authors, web pages or publishers. All other brands or product names used in this Guide are the trade names or registered trademarks of their respective owners. ~the author, Dan Tousignant

Agile Contracting	The first lesson we learnt in contracting out agile software development (or anything else for that matter), is to align objectives of the supplier and the customer. It is highly desirable to align supplier success with customer success.
	The key here is to define the product vision and what must be achieved; foster shared ownership of the goals by treating your supplier as a partner; and consider offering the supplier incentives for meeting key business performance indicators that require partnership with you.
	http://blog.scrumup.com/2012/11/top-ten-reads-on-agile-contracts.html
	http://agilesoftwaredevelopment.com/blog/peterstev/10-agile-contracts
Agile Discovery (New 2015)	During the Discovery Phase, designers need to work with the business analyst to capture and define business requirements. This is done by facilitating workshops and interviewing key stakeholders. A lot of sketching, note taking, brainstorming and discussion happen at this stage in order to effectively visualize the early thinking on look and feel, layout and interaction design.

	It is also worth noting that defining the business model is an evolutionary process. At the end of the discovery and design phases the value proposition needs to map back to real user personas, partnerships, activities, a cost structure, solid business KPIs and have a business mission statement should all clearly defined. http://www.thoughtworks.com/insights/blog/providing-just-enough-design-can-make-agile-software-delivery-more-successful
Architectural Spike (New 2015)	XP does this while the initial Planning Game is in process. It's not an iteration - it might be longer or shorter, we don't know yet. What it's about is exploring solution elements that seem relevant to the as yet limited knowledge we have about the problem domain, choosing a System Metaphor, putting enough of our build, dbms, and source control tools in place to be able to begin controlled work, and then proceeding until we have something that runs and can be iterated. https://www.linkedin.com/pulse/agile-architecting-practice-architecture-spike-erik-philippus
Active Listening (New 2015)	Active listening is a communication technique used in counselling, training and conflict resolution, which requires the listener to feed back what they hear to the speaker, by way of re-stating or paraphrasing what they have heard in their own words, to confirm what they have heard and moreover, to confirm the understanding of both parties https://en.wikipedia.org/wiki/Active_listening
Adaptive Leadership	Adaptive leadership focuses on team management, from building self-organizing teams to developing a servant leadership style. It is both more difficult, and ultimately more rewarding than managing tasks. In an agile enterprise the people take care of the tasks and the leader engages the people. The facilitative leader works on things like building self-organizing teams, a trusting and respectful environment, collaboration, participatory decision making, and developing appropriate empowerment guidelines (for an excellent discussion of empowerment, see Chapters 6 & 7 in (Appelo, 2011)). http://www.thoughtworks.com/sites/www.thoughtworks.com/files/files/adaptive-leadership-accelerating-enterprise-agility-jim-highsmith-thoughtworks.pdf Appelo, J. (2011). Management 3.0: Leading Agile Developers, Developing Agile Leaders. Upper Saddle River, JN: Addison-Wesley.
Adaptability, Three Components	Adaptability has three components—**product, process, and people.** You need to have a gung-ho agile team with the right attitude about change. You need processes and practices that allow the team to adapt to circumstances. And you *need* high quality code with automated tests. You can have pristine code and a non-agile team and change will be difficult. All three are required to have

Agile Certification Study Guide – Glossary

	an agile, adaptable environment. http://searchsoftwarequality.techtarget.com/feature/Adaptation-in-project-management-through-agile
Affinity Estimating	A facilitated process where team members for sequence the product backlog from smallest to largest user story, then the rest of the team validates and finally the user stories are group by a sizing method such as t-shirt size or Fibonacci sequence. http://www.gettingagile.com/2008/07/04/affinity-estimating-a-how-to/ http://www.agilebok.org/index.php?title=Affinity_Estimating
Agile Product Lifecycle	**The Five Agile plans:** 1 – Product Vision: Yearly By Product Owner 2 – Product Roadmap: Bi-Yearly By The Product Owner 3 – Release Plan: Quarterly By The Product Owner And Teams 4 – Iteration Plan: Bi-Weekly By The Teams 5 – Daily Plan (Scrum): Daily By Individual (EXAM TIP: different authors reference different time periods for updating these plans) http://www.romanpichler.com/blog/
Agile Modeling (AM)	AM is a collection of values, principles, and practices for modeling software that can be applied on a software development project in an effective and light-weight manner. Values: 1. Communication 2. Simplicity 3. Feedback 4. Courage 5. Humility Principles/Practices: - <u>Active Stakeholder Participation</u>. Stakeholders should provide information in a timely manner, make decisions in a timely manner, and be as actively involved in the development process through the use of <u>inclusive tools and techniques.</u> - <u>Architecture Envisioning.</u> At the beginning of an agile project you will need to do some initial, high-level architectural modeling to identify a viable technical strategy for your solution.

Agile Certification Study Guide – Glossary

	- **Document Continuously.** Write deliverable documentation throughout the lifecycle in parallel to the creation of the rest of the solution. - **Document Late.** Write deliverable documentation as late as possible, avoiding speculative ideas that are likely to change in favor of stable information. - **Executable Specifications.** Specify requirements in the form of executable "customer tests", and your design as executable developer tests, instead of non-executable "static" documentation. - **Iteration Modeling.** At the beginning of each iteration you will do a bit of modeling as part of your iteration planning activities. - **Just Barely Good Enough (JBGE) artifacts.** A model or document needs to be sufficient for the situation at hand and no more. - **Look Ahead Modeling.** Sometimes requirements that are nearing the top of your priority stack are fairly complex, motivating you to invest some effort to explore them before they're popped off the top of the work item stack so as to reduce overall risk. - **Model Storming.** Throughout an iteration you will model storm on a just-in-time (JIT) basis for a few minutes to explore the details behind a requirement or to think through a design issue. - **Multiple Models.** Each type of model has its strengths and weaknesses. An effective developer will need a range of models in their intellectual toolkit enabling them to apply the right model in the most appropriate manner for the situation at hand. - **Prioritized Requirements.** Agile teams implement requirements in priority order, as defined by their stakeholders, so as to provide the greatest return on investment (ROI) possible. - **Requirements Envisioning.** At the beginning of an agile project you will need to invest some time to identify the scope of the project and to create the initial prioritized stack of requirements. - **Single Source Information.** Strive to capture information in one place and one place only. - **Test-Driven Design (TDD).** Write a single test, either at the requirements or design level, and then just enough code to fulfill that test. TDD is a JIT approach to detailed requirements specification and a confirmatory approach to testing. http://www.agilemodeling.com/
Agile Scaling Model	Agile scaling factors are: 1. Team size 2. Geographical distribution 3. Regulatory compliance 4. Domain complexity 5. Organizational distribution 6. Technical complexity 7. Organizational complexity 8. Enterprise discipline

Agile Certification Study Guide – Glossary

	http://www.agilealliance.org/files/session_pdfs/Agile%20Scaling%20Model.pdf
Agile Triangle	The three dimensions critical to Agile performance measurement : - Value, - Quality - Constraints (cost, schedule, scope). Also, simplified to be: - Value - Technical debt - Cost http://jimhighsmith.com/beyond-scope-schedule-and-cost-the-agile-triangle/
APM Delivery Framework	However, if the business objective is reliable innovation, then the process framework must be organic, flexible, and easy to adapt. The APM process framework supports this second business objective through the five phases of: - Envision - Speculate - Explore - Adapt - Close http://www.informit.com/articles/article.aspx?p=174660&seqNum=4
Asynchronous Builds	When you use the integration script discussed earlier, you're using synchronous integration—you're confirming that the build and tests succeed before moving on to your next task. If the build is too slow, synchronous integration becomes untenable. (For me, 20 or 30 minutes is too slow.) In this case, you can use asynchronous integration instead. Rather than waiting for the build to complete, start your next task immediately after starting the build, without waiting for the build and tests to succeed. The biggest problem with asynchronous integration is that it tends to result in broken builds. If you check in code that doesn't work, you have to interrupt what you're doing when the build breaks half an hour or an hour later. If anyone else checked out that code in the meantime, their build won't work either. If the pair that broke the build has gone home or to lunch, someone else has to clean up the mess. In practice, the desire to keep working on the task at hand often overrides the need to fix the build. If you have a very slow build, asynchronous integration may be your only option. If you must use this, a continuous integration server is the best way to do so. It will keep track of what to build and automatically notify you when the build has finished.

	http://jamesshore.com/Agile-Book/continuous_integration.html
Backlog Grooming/ Refinement (new 2015)	Product Backlog refinement is the act of adding detail, estimates, and order to items in the Product Backlog. This is an ongoing process in which the Product Owner and the Development Team collaborate on the details of Product Backlog items. During Product Backlog refinement, items are reviewed and revised. The Scrum Team decides how and when refinement is done. Refinement usually consumes no more than 10% of the capacity of the Development Team. However, Product Backlog items can be updated at any time by the Product Owner or at the Product Owner's discretion. http://www.scrumguides.org/scrum-guide.html
Brainstorming (New 2015)	A brainstorming session is a tool to generate ideas from a selected audience to solve a problem or stimulate creativity. These meetings are used for solving a process problem, inventing new products or product innovation, solving inter-group communication problems, improving customer service, budgeting exercises, project scheduling, etc. Using these tools can help discussion facilitators and Project Managers with alternative approaches for creative idea generation meetings (aka Brainstorming Sessions). They are particularly useful when previous meetings have gone afoul, are not as effective as they could be or productivity during these exercises is less than it should be. http://www.projectconnections.com/templates/detail/brainstorming-meeting-techniques.html
Burn Down Charts	

	X-Axis	The project/iteration timeline
	Y-Axis	The work that needs to be completed for the project. The time estimates for the work remaining will be represented by this axis.
	Project Start Point	This is the farthest point to the left of the chart and occurs at day 0 of the project/iteration.
	Project End Point	This is the point that is farthest to the right of the chart and occurs on the predicted last day of the project/iteration
	Ideal Work Remaining Line	This is a straight line that connects the start point to the end point. At the start point, the ideal line shows the sum of the estimates for all the tasks (work) that needs to be completed. At the end point, the ideal line intercepts the x-axis showing

		that there is no work left to be completed.
	Actual Work Remaining Line	This shows the actual work remaining. At the start point, the actual work remaining is the same as the ideal work remaining but as time progresses; the actual work line fluctuates above and below the ideal line depending on how effective the team is. In general, a new point is added to this line each day of the project. Each day, the sum of the time estimates for work that was recently completed is subtracted from the last point in the line to determine the next point.
	Actual Work Line is above the Ideal Work Line	If the actual work line is above the ideal work line, it means that there is more work left than originally predicted and the project is behind schedule.
	Actual Work Line is below the Ideal Work Line	If the actual work line is below the ideal work line, it means that there is less work left than originally predicted and the project is ahead of schedule.

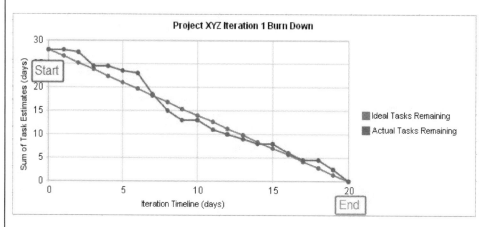

http://en.wikipedia.org/wiki/Burn_down_chart

Burndown Bar Charts	The typical Scrum release burndown chart shows a single value--the net change in the amount of work remaining. • As tasks are completed, the top of the bar is lowered. • When tasks are added to the original set, the bottom of the bar is lowered. • When tasks are removed from the original set, the bottom of the bar is raised. • When the amount of work involved in a task changes, the top of the bar moves up or down.

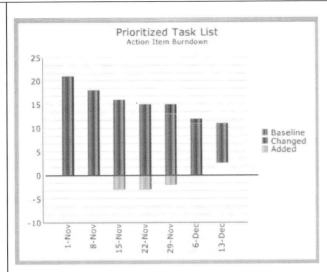

http://www.snyders.us/burndown-bar-tracking.htm

Burn Up Charts	The amount of accepted work (that work which has been completed, tested, and met acceptance criteria) is graphed in a burnup chart. The amount of work in an accepted state starts at 0 and continues to grow until it reaches 100% accepted at the end of the Iteration. 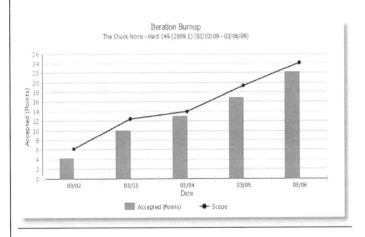 http://www.rallydev.com/help/progress-reports
Chartering in Agile	1. **Vision:** The vision defines the "Why" of the project. This is the higher purpose, or the reason for the project's existence. 2. **Mission:** This is the "What" of the project and it states what will be done in the project to achieve its higher purpose. 3. **Success Criteria:** The success criteria are management tests that describe effects outside of the solution itself. http://www.agilebok.org/index.php?title=Elements_of_a_Project_Charter_for_an

Agile Certification Study Guide –Glossary

	Agile Project
Collaboration	Collaboration is the basis for bringing together the knowledge, experience and skills of multiple team members to contribute to the development of a new product more effectively than individual team members performing their narrow tasks in support of product development. As such collaboration is the basis for concepts such as concurrent engineering or integrated product development. Effective collaboration requires actions on multiple fronts: • Early involvement and the availability of resources to effectively collaborate • A culture that encourages teamwork, cooperation and collaboration • Effective teamwork and team member cooperation • Defined team member responsibilities based on collaboration • A defined product development process based on early sharing of information and collaborating • Collocation or virtual collocation • Collaboration technology (EXAM TIP: There was specifically a question about the difference between coordination and collaboration) http://www.agilebok.org/index.php?title=Collaboration
Collaboration Games	Agile games are activities focused on teaching, demonstrating, and improving Agile and organizational effectiveness using game theory. Using games, we can model complex or time-consuming processes and systems, examine why they work (or don't work), look for improvements, and teach others how to benefit from them. Games can model just the core of a process or model, leaving out unimportant factors. They can involve collaboration, brainstorming, comparing variants, and of course retrospectives. http://agilegamesnewengland.com/
Collective Code Ownership	The way this works is for each developer to create unit tests for their code as it is developed. All code that is released into the source code repository includes unit tests that run at 100%. Code that is added, bugs as they are fixed, and old functionality as it is changed will be covered by automated testing. Now you can rely on the test suite to watch dog your entire code repository. Before any code is released it must pass the entire test suite at 100%. Once this is in place anyone can make a change to any method of any class and release it to the code repository as needed. When combined with frequent integration developers rarely even notice a class has been extended or repaired. http://www.extremeprogramming.org/rules/collective.html

Agile Certification Study Guide – Glossary

Collocated And Or Distributed Teams	Collocated Agile is a model in which projects execute the Agile Methodology with teams located in a single room. The methodology requires that the complete team be in close proximity to each other to improve coordination between the members. Collocated Agile teams have proven that the real power of project success lies not in administration, but in the acumen, chemistry, loyalty, and dedication between the collocated teams. Distributed Agile, as the name implies, is a model in which projects execute an Agile Methodology with teams that are distributed across multiple geographies. http://www.continuousagile.com/unblock/team_options.html
Compliance	Doing agile development in a relatively inelastic environment, where policies and procedures are virtually unchangeable, creates an impedance mismatch between the agile team and its host organization. Our experience on a variety of embedded Java projects has shed some light on the costs of complying (or failing to comply), where we trialed everything from "refusal to comply" to "full compliance". Regardless of approach, there was always an associated cost, whether in redrafting documents, reducing functionality, spending time in meetings, losing focus on deliverables, or deteriorating morale. In keeping with the Agile philosophy, when our efforts were failing, we refactored our approach to managing the project in an effort to minimize the costs of compliance without adopting more risk to ensure success. In the end we were faced with the questions "How did we fare in the end? Which costs were worth bearing? Was it all worth it?" http://sais.aisnet.org/2008/1AMishraWeistroffer.pdf
Conflict Resolution	In some cases, conflicts between members of a team or a department may be so hard to diffuse that only the leader can actually help resolve it. The manager should be equipped with the right things to say and do in order to utilize conflict resolution strategies to solve the conflicts between the members of the team so that it will not affect the day to day operations of the business. It can be quite challenging to handle such conflicts even with a seasoned manager in charge. However, there are some tips that you can look into so you can properly solve the major concerns of each party and hopefully come up with a compromise between them. If conflicts remain unresolved, there is a greater chance of decreasing productivity in the workplace and declining proficiency. http://www.agilebok.org/index.php?title=Conflict_Resolution http://www.ameaningfulexistence.com/2010/03/22/know-these-five-causes-of-conflict/ http://www.agilecoach.net/wp-content/uploads/2009/10/Conflict-Resolution-Diagram.pdf

Conflict Levels

Level 1: Problem to Solve

We all know what conflict at level 1 feels like. Everyday frustrations and aggravations make up this level, and we experience conflicts as they rise and fall and come and go. At this level, people have different opinions, misunderstanding may have happened, conflicting goals or values may exist, and team members likely feel anxious about the conflict in the air.

When in level 1, the team remains focused on determining what's awry and how to fix it. Information flows freely, and collaboration is alive. Team members use words that are clear, specific, and factual. The language abides in the here and now, not in talking about the past. Team members check in with one another if they think a miscommunication has just happened. You will probably notice that team members seem optimistic, moving through the conflict. It's not comfortable, but it's not emotionally charged, either. Think of level 1 as the level of constructive disagreement that characterizes high-performing teams.

Level 2: Disagreement

At level 2, self-protection becomes as important as solving the problem. Team members distance themselves from one another to ensure they come out OK in the end or to establish a position for compromise they assume will come. They may talk offline with other team members to test strategies or seek advice and support. At this level, good-natured joking moves toward the half-joking barb. Nastiness gets a sugarcoating but still comes across as bitter. Yet, people aren't hostile, just wary. Their language reflects this as their words move from the specific to the general. Fortifying their walls, they don't share all they know about the issues. Facts play second fiddle to interpretations and create confusion about what's really happening.

Level 3: Contest

At level 3, the aim is to win. A compounding effect occurs as prior conflicts and problems remain unresolved. Often, multiple issues cluster into larger issues or create a "cause." Factions emerge in this fertile ground from which misunderstandings and power politics arise. In an agile team, this may happen subtly, because a hallmark of working agile is the feeling that we are all in this together. But it does happen.

People begin to align themselves with one side or the other. Emotions become tools used to "win" supporters for one's position. Problems and people become synonymous, opening people up to attack. As team members pay attention to building their cases, their language becomes dis-torted. They make overgeneralizations: "He always forgets to check in his code" or "You never listen to what I have to say." They talk about the other side in presumptions: "I know what they think, but they are ignoring the real issue." Views of themselves as benevolent and others as tarnished become magnified: "I am always the one to compromise for the good of the team" or "I have everyone's best interest at heart" or "They are intentionally ignoring what the customer is really saying." Discussion becomes either/or and blaming flourishes. In this combative environment, talk of peace may meet resistance. People may not be ready to move beyond blaming.

Level 4: Crusade

At level 4, resolving the situation isn't good enough. Team members believe the people on the "other side" of the issues will not change. They may believe the only option is to remove the others from the team or get removed from the team themselves. Factions become entrenched and can even solidify into a pseudo-organizational structure within the team. Identifying with a faction can overshadow identifying with the team as a whole so the team's identity gets trounced. People and positions are seen as one, opening up people to attack for their affiliations rather than their ideas. These attacks come in the form of language rife with ideology and principles, which becomes the focus of conversation, rather than specific issues and facts. The overall attitude is righteous and punitive.

Level 5: World War

"Destroy!" rings out the battle cry at level 5. It's not enough that one wins; others must lose. "We must make sure this horrible situation does not happen again!" Only one option at level 5 exists: to separate the combatants (aka team members) so that they don't hurt one another. No constructive outcome can be had.
http://agile.dzone.com/articles/agile-managing-conflict

Conflict Types	**#1 - Lack of Role Clarity** The project manager is responsible for assigning tasks to each project team member. In addition, they often assume that team members understand what is being asked of them. This assumption can be incorrect, leading to team members being unclear on what needs to be accomplished. A good project manager takes the time to explain the tasks, their expectations and timeframes around completion. **#2 - Difference in Prioritizing Tasks** Just because the project manager thinks the task is a milestone, the team member completing the task may not. Team members may be working simultaneously on multiple projects and cannot differentiate the priority of one project's tasks from another. The project manager should try to explain the importance of the overall project to the **#3 - Working in Silos** Often, project team members work independently. They may work remotely or in a different location from other project team members. Conflict arises when team members are not aware of what others are doing and are not communicating with one another. The project manager needs to bring the team together to discuss project status and barriers to getting the project completed promptly. If team members working in silos can envision how they are a part of the bigger picture, they will be more motivated and feel like a part of the team. **#4 - Lack of Communication**

Agile Certification Study Guide – Glossary

	Project managers must foster a clear line of communication between project team members. In order to minimize duplication of efforts, the project manager should communicate expectations to all team members. The project manager needs to be easily accessible to project team members at all times during the project. If team members cannot reach their project manager or other team members, they may spin in circles needlessly. **# 5 - Waiting on Completion of Task Dependencies** Some tasks cannot be started until other tasks are completed. Team members need to understand the impact of their role on others. For example if one team member is responsible for ordering equipment and another for installing the equipment, one task is dependent on the other. Conflict can occur if the first team member is delayed in completing their tasks. http://www.brighthub.com/office/project-management/articles/95971.aspx
Container	A container is a closed space where things can get done, regardless of the overall complexity of the problem. In the case of Scrum, a container is a Sprint, an iteration. We put people with all the skills needed to solve the problem in the container. We put the highest value problems to be solved into the container. Then we protect the container from any outside disturbances while the people attempt to bring the problem to a solution. We control the container by time-boxing the length of time that we allow the problem to be worked on. We let the people select problems of a size that can be brought to fruition during the time-box. At the end of the time-box, we open the container and inspect the results. We then reset the container (adaptation) for the next time-box. By frequently replanning and shifting our work, we are able to optimize value. http://kenschwaber.wordpress.com/2010/06/10/waterfall-leankanban-and-scrum-2/
Continuous Integration 	Continuous Integration (CI) involves producing a clean build of the system several times per day. Agile teams typically configure CI to include automated compilation, unit test execution, and source control integration. Sometimes CI also includes automatically running automated acceptance tests. **Continuous Integration Technique, Tools, and Policy** There are several specific practices that CI seems to require to work well. On his site, Martin Fowler provides a long, detailed description of what Continuous Integration is and how to make it work. One popular CI rule states that programmers never leave anything unintegrated at the end of the day. The build should never spend the night in a broken state. This imposes some task planning discipline on programming teams. Furthermore, if the team's rule is that whoever breaks the build at check-in has to fix it again, there is a natural incentive to check code in frequently during the day. **Benefits of Continuous Integration** When CI works well, it helps the code stay robust enough that customers and other stakeholders can play with the code whenever they like. This speeds the flow of development work overall; as Fowler points out, it has a very different feel

	to it. It also encourages more feedback between programmers and customers, which helps the team get things right before iteration deadlines. Like refactoring, continuous integration works well if you have an exhaustive suite of automated unit tests that ensure that you are not committing buggy code. http://www.versionone.com/Agile101/Continuous_Integration.asp http://www.martinfowler.com/articles/continuousIntegration.html
Control Limits (New 2015)	Control limits, also known as natural process limits, are horizontal lines drawn on a statistical process control chart, usually at a distance of ±3 standard deviations of the plotted statistic from the statistic's mean. Used in Agile Control charts https://www.youtube.com/watch?v=JNZmNWOMv2w
Cumulative Flow Diagrams	The Cumulative Flow diagram is very similar to a Burnup Chart. It shows how much of our work (i.e., the effort associated with User Stories) is in different states, such as Completed or In Progress. The Total and Completed curves shows the Release scope and Burnup of completed work, while the In Progress curve shows how much work is associated with Stories currently in development. The primary difference from the standard Burnup and Cumulative Flow diagrams is that the latter shows how much of the work is currently in progress. *FDD Cumulative Flow Diagram chart showing Features (0-700) vs Time (weeks 1-29), with areas for Not Started, Started, and Completed.* http://www.cprime.com/store/scrum_and_agile_essentials/cumulative_flow_diagram_burnup_chart_.html
Customer-Valued Prioritization	Agile development is about the frequent delivery of high-value, working software to the customer/user community. Doing so requires the prioritization of user stories and the continuous monitoring of the prioritized story backlog. The primary driver for prioritization is customer value. However, it is insufficient to simply say that the highest-value stories are the highest priority. Product owners must also factor in the cost of development. An extremely valuable feature quickly loses its luster when it is also extremely costly to implement. Additionally, there are other secondary drivers such as risk and uncertainty. These should be resolved early.

Agile Certification Study Guide – Glossary

	There may also be experimental stories that are worth developing early to find out whether customers see value in further development along those lines. There may be other prioritization drivers, but business value should always be foremost. 1. Complete the high-value, high-risk stories first if the cost is justified. 2. Complete the high-value, low-risk stories next if the cost is justified. 3. Complete the lower-value, low-risk stories next. 4. Avoid low-value, high-risk stories. http://www.scribd.com/doc/111905434/Agile-Analytics-a-Value-Driven-Approach-Ken-Collier
Cycle Time	Cycle time for software development is measured in the number of days needed between feature specification and production delivery. This is called: Software In Process (SIP). A shorter cycle indicates a healthier project. A Lean project that deploys to production every 2-weeks has a SIP of 10 working days. Some Lean projects even deploy nightly. http://www.agilebok.org/index.php?title=Cycle_Time
Daily Scrums/Daily Stand-Ups 	Time-boxed 15 minutes meetings whose purpose is the providing of a concise team status. Scrum has each team member asked by the ScrumMaster the following questions: • What did you do yesterday? • What are you doing today? • Are there any impediments that need resolution? (EXAM TIP: I had a couple of questions relating to the purpose of this meeting) http://www.mountaingoatsoftware.com/scrum/daily-scrum
Declaration Of Interdependence	The **PM Declaration of interdependence** is a set of six management principles initially intended for project managers of Agile Software Development projects We are a community of project leaders that are highly successful at delivering results. To achieve these results: • We **increase return on investment** by making continuous flow of value our focus. • We **deliver reliable results** by engaging customers in frequent interactions and shared ownership. • We **expect uncertainty** and manage for it through iterations, anticipation, and adaptation. • We **unleash creativity and innovation** by recognizing that individuals are the ultimate source of value, and creating an environment where they can make a difference. • We **boost performance** through group accountability for results and shared responsibility for team effectiveness.

	- We **improve effectiveness and reliability** through situationally specific strategies, processes and practices. http://pmdoi.org/
D.E.E.P.	- **Detailed Appropriately.** User stories on the product backlog that will be done soon need to be sufficiently well understood that they can be completed in the coming sprint. Stories that will not be developed for awhile should be described with less detail. - **Estimated.** The product backlog is more than a list of all work to be done; it is also a useful planning tool. Because items further down the backlog are not as well understood (yet), the estimates associated with them will be less precise than estimates given items at the top. - **Emergent.** A product backlog is not static. It will change over time. As more is learned, user stories on the product backlog will be added, removed, or reprioritized. - **Prioritized.** The product backlog should be sorted with the most valuable items at the top and the least valuable at the bottom. By always working in priority order, the team is able to maximize the value of the product or system being developed. http://www.mountaingoatsoftware.com/blog/make-the-product-backlog-deep
Definition Of Done	Definition of Done (DoD) is a simple list of activities (writing code, coding comments, unit testing, integration testing, release notes, design documents, etc.) that add verifiable/demonstrable value to the product. Focusing on value-added steps allows the team to focus on what must be completed in order to build software while eliminating wasteful activities that only complicate software development efforts. Note – the DoD is defined by the Product Owner and committed to by the team. http://www.scrumalliance.org/articles/105-what-is-definition-of-done-dod
Defect Rate (New 2015)	The defect detection rate is the amount of defects detected per sprint. Assuming that developers produce defects at a more or less constant rate, it is correlated with the velocity; the more story points are delivered, the more defects should be found and fixed as well. Teams tend to be pretty consistent in the quality of the software they deliver, so a drop in velocity combined with a rise in the defect detection rate should trigger the alarm. Something's cooking and you need to find out what it is. My personal opinion is that a lower defect detection rate isn't necessarily better than a higher one. A defect more found in one of the development and test environments is a defect less that makes it into production. From that perspective, you could support the statement, the more defects the better. http://theagileprojectmanager.blogspot.com/2013/05/agile-metrics.html

Agile Certification Study Guide – Glossary

Discounted Pay-Back Period	A capital budgeting procedure used to determine the profitability of a project. In contrast to an NPV analysis, which provides the overall value of a project, a discounted payback period gives the number of years it takes to break even from undertaking the initial expenditure. Future cash flows are considered are discounted to time "zero." This procedure is similar to a payback period; however, the discounted payback period measures how long it take for the initial cash outflow to be paid back, including the time value of money. http://accountingexplained.com/managerial/capital-budgeting/discounted-payback-period
Dreyfus Model of Skill Acquisition (New 2015)	**1. Novice** - "rigid adherence to taught rules or plans" - no exercise of "discretionary judgment" **2. Advanced beginner** - limited "situational perception" - all aspects of work treated separately with equal importance **3. Competent** - "coping with crowdedness" (multiple activities, accumulation of information) - some perception of actions in relation to goals - deliberate planning - formulates routines **4. Proficient** - holistic view of situation - prioritizes importance of aspects - "perceives deviations from the normal pattern" - employs maxims for guidance, with meanings that adapt to the situation at hand **5. Expert** - transcends reliance on rules, guidelines, and maxims - "intuitive grasp of situations based on deep, tacit understanding" - has "vision of what is possible" - uses "analytical approaches" in new situations or in case of problems https://en.wikipedia.org/wiki/Dreyfus_model_of_skill_acquisition
DRY	Software must be written expecting for future change. Principles like **D**on't **R**epeat **Y**ourself (DRY) are used to facilitate this. In agile development, changes to the software specifications are welcome even in late stages of development. As clients get more hands-on time with iterative builds of the software, they may be able to better communicate their needs. Also, changes in the market or

	company structure might dictate changes in the software specifications. Agile development is designed to accommodate these late changes. http://ruby.about.com/od/rubyonrails/a/agile.htm
Earned Value Management (EVM)	Agile EVM is now all about executing the project and tracking the accumulated EV according to the simple earning rule. Because Agile EVM has been evolving for many years the following practices are well-established: • EV is accumulated at fixed time intervals (i.e. Timebox, Iteration or Sprint) of 1–4 weeks; • PV and EV is graphically tracked & extrapolated as remaining value in a Release Burndown Chart as shown in figure 6; • Rather than a S-shaped curve the PV in Agile EVM is a straight line because an Agile project has no distinct phases and corresponding variances in the rate of value delivery; • The EV in Story Points done in one fixed time interval is known as the Velocity of a team; • In Agile scope change is embraced and the amount of added (removed) scope in Story Points is added (removed) to the Velocity or Scope Floor. The latter is shown in Figure 6 where several scope increases have lowered the Scope Floor below the x-axis. The advantage of using a Scope Floor is that any scope changes can easily be separated from Velocity variances; • The intersection between the Remaining Value and Scope Floor lines indicates the expected release date and the corresponding Remaining Budget. (EXAM TIP –There was a question on the exam that asked when was the best time to measure EVM) http://en.wikipedia.org/wiki/Earned_value_management#Agile_EVM
Emotional Intelligence	**Self Awareness** Awareness of one's own feelings and the ability to recognize and manage these feelings in a way which one feels that one can control. This factor includes a degree of self-belief in one's ability to manage one's emotions and to control their impact in a work environment. **Emotional Resilience** The ability to perform consistently in a range of situations under pressure and to adapt one's behavior appropriately. The facility to balance the needs and concerns of the individuals involved. The ability to retain focus on a course of action or need for results in the face of personal challenge or criticism. **Motivation** The drive and energy to achieve clear results and make an impact: and to balance both short and long term goals with an ability to pursue demanding goals in the face of rejection or questioning. **Interpersonal Sensitivity** The facility to be aware of, and take account of, the needs and perceptions of

others when arriving at decisions and proposing solutions to problems and challenges. The ability to build from this awareness and achieve 'buy-in' to decisions and ideas for action.

Influence
The ability to persuade others to change a viewpoint based on the understanding of the position and the recognition of the need to listen to this perspective and provide a rationale for change.

Intuitiveness
The ability to arrive at clear decisions and drive their implementation when presented with incomplete or ambiguous information using both rational and 'emotional' or insightful perceptions of key issues and implications.

Conscientiousness
The ability to display clear commitment to a course of action in the face of challenge and to match 'words and deeds'; in encouraging others to support the chosen direction. The personal commitment to pursuing an ethical solution to a difficult business issue or problem.
http://businessagile.blogspot.com/2005/02/emotional-intelligence-key-element-of.html

Empirical Process Control

Empirical Process Control

The word "empirical" denotes information gained by means of observation, experience, or experiment. The empirical process control constitutes a continuous cycle of inspecting the process for correct operation and results and adapting the process as needed. There are three legs that hold up every implementation of empirical process control: **transparency, inspection, and adaptation.** The first leg is transparency. It means that those aspects of the process that affect the outcome must be visible to those controlling the process. The second leg is inspection. The various aspects of the process must be inspected frequently enough that unacceptable variances in the process can be detected. The third leg of empirical process control is adaptation. The process or the material being processed must be adjusted if one or more aspects of the process are outside acceptable limits and the resulting product will be unacceptable.

http://www.scrumstudy.com/scrum-empirical-process-control.asp

Escaped Defects

Escaped Defects are those defects reported by the Customer that have escaped all software quality processes are represented in this metric. Escaping defects should then be treated as ranked backlog work items, along with other project work items. They should be prioritized high enough to resolve them within the next sprint or two and not accumulate a growing backlog. Watch the defect backlog as part of the project metrics. A growing defect backlog is a key indicator that the team is taking on more new work than it can handle. It may also be a key indicator that the team is operating as a "mini-waterfall" project, rather than a agile project, requiring more collaboration between Dev and

	Quality Engineers and early testing. Drop the number of new items the team works on until the escaping defects are well managed or eliminated. http://www.agilebok.org/index.php?title=Escaped_Defects
Exploratory Testing	Exploratory Testing is a technique for finding surprising defects. Testers use their training, experience, and intuition to form hypotheses about where defects are likely to be lurking in the software, then they use a fast feedback loop to iteratively generate, execute, and refine test plans that expose those defects. It appears similar to ad-hoc testing to an untrained observer, but it's far more rigorous. Some teams use exploratory testing to check the quality of their software. After a story's been coded, the testers do some exploratory testing, the team fixes bugs, and repeat. Once the testers don't find any more bugs, the story is done. (EXAM TIP: This is also referred to as testing performed between Done and Done-Done) http://jamesshore.com/Blog/Alternatives-to-Acceptance-Testing.html

Extreme Programming	Extreme Programming (or XP) is a set of values, principles and practices for rapidly developing high-quality software that provides the highest value for the customer in the fastest way possible. XP is extreme in the sense that it takes 12 well-known software development "best practices" to their logical extremes.
	
	The 12 core practices of XP are:
	1) **The Planning Game**: Business and development cooperate to produce the maximum business value as rapidly as possible. The planning game happens at various scales, but the basic rules are always the same:
	   a) Business comes up with a list of desired features for the system. Each feature is written out as a **User Story**, which gives the feature a name, and describes in broad strokes what is required. User stories are typically written on 4x6 cards.
	   b) Development estimates how much effort each story will take, and how much effort the team can produce in a given time interval (the iteration).
	   c) Business then decides which stories to implement in what order, as well as when and how often to produce a production releases of the system.
	2) **Small Releases**: Start with the smallest useful feature set. Release early and often, adding a few features each time.
	3) **System Metaphor**: Each project has an organizing metaphor, which provides an easy to remember naming convention.
	4) **Simple Design**: Always use the simplest possible design that gets the job done. The requirements will change tomorrow, so only do what's needed to meet today's requirements.
	5) **Continuous Testing**: Before programmers add a feature, they write a test for it. When the suite runs, the job is done. Tests in XP come in two basic flavors.
	   a) **Unit Tests** are automated tests written by the developers to test

functionality as they write it. Each unit test typically tests only a single class, or a small cluster of classes. Unit tests are typically written using a unit testing framework, such as JUnit.

  b) **Acceptance Tests** (also known as **Functional Tests**) are specified by the customer to test that the overall system is functioning as specified. Acceptance tests typically test the entire system, or some large chunk of it. When all the acceptance tests pass for a given user story, that story is considered complete. At the very least, an acceptance test could consist of a script of user interface actions and expected results that a human can run. Ideally acceptance tests should be automated, either using the unit testing framework, or a separate acceptance testing framework.

6) **Refactoring**: Refactor out any duplicate code generated in a coding session. You can do this with confidence that you didn't break anything because you have the tests.

7) **Pair Programming**: All production code is written by two programmers sitting at one machine. Essentially, all code is reviewed as it is written.

8) **Collective Code Ownership**: No single person "owns" a module. Any developer is expect to be able to work on any part of the codebase at any time.

9) **Continuous Integration**: All changes are integrated into the codebase at least daily. The tests have to run 100% both before and after integration.

10) **40-Hour Work Week**: Programmers go home on time. In crunch mode, up to one week of overtime is allowed. But multiple consecutive weeks of overtime are treated as a sign that something is very wrong with the process.

11) **On-site Customer**: Development team has continuous access to a real live customer, that is, someone who will actually be using the system. For commercial software with lots of customers, a customer proxy (usually the product manager) is used instead.

12) **Coding Standards**: Everyone codes to the same standards. Ideally, you shouldn't be able to tell by looking at it who on the team has touched a specific piece of code.

(EXAM TIP: Lots of questions about Lean, XP and Scrum principles)

http://www.jera.com/techinfo/xpfaq.html

Feature Breakdown Structure (FBS)	During detailed planning, agile development favors a feature breakdown structure (FBS) approach instead of the work breakdown structure (WBS) used in waterfall development approaches. Feature breakdown structures are advantageous for a few reasons: 1. They allow communication between the customer and the development team in terms both can understand. 2. They allow the customer to prioritize the team's work based on business value. 3. They allow tracking of work against the actual business value produced. It is acceptable to start out with features that are large and then break them out into smaller features over time. This allows the customer to keep from diving in to

Agile Certification Study Guide –Glossary

	too much detail until that detail is needed to help facilitate actual design and delivery. http://www.versionone.com/agile-101/agile-feature-estimation/
Fishbone diagram **(New 2015)**	The fishbone diagram identifies many possible causes for an effect or problem. It can be used to structure a brainstorming session. It immediately sorts ideas into useful categories. http://asq.org/learn-about-quality/cause-analysis-tools/overview/fishbone.html
The Five Whys **(New 2015)**	A technique for identifying the exact root cause of a problem to determine the appropriate solution. Made popular in the 70s by the Toyota Production System, the 5 whys is a flexible problem solving technique. http://businessanalystlearnings.com/ba-techniques/2013/2/5/root-cause-analysis-the-5-whys-technique
Fractional Assignments	All of the team members should sit with the team full-time and give the project their complete attention. This particularly applies to customers, who are often surprised at the level of involvement XP requires of them. Some organizations like to assign people to multiple projects simultaneously. This fractional assignment is particularly common in matrix-managed organizations. (If team members have two managers, one for their project and one for their function, you're probably in a matrixed organization.) Fractional assignment is dreadfully counterproductive. If your company practices fractional assignment, I have some good news. You can instantly improve productivity by reassigning people to only one project at a time. Fractional assignment is dreadfully counterproductive: fractional workers don't bond with their teams; they often aren't around to hear conversations and answer questions and they must task switch, which incurs a significant hidden penalty. "The minimum penalty is 15 percent... Fragmented knowledge workers may look busy, but a lot of their busyness is just thrashing." [DeMarco 2002] (p.19-20) That's not to say that everyone needs to work with the team for the entire duration of the project. You can bring someone in to consult on a problem temporarily. However, while she works with the team, she should be fully engaged and available. http://jamesshore.com/Agile-Book/the_xp_team.html Tom DeMarco and Barry Boehm. 2002. The Agile Methods Fray. *Computer* 35, 6 (June 2002)

Frequent Verification And Validation	The terms Verification and Validation are commonly used in software engineering to mean two different types of analysis. The usual definitions are: **Validation**: Are we building the right system? **Verification**: Are we building the system right? In other words, validation is concerned with checking that the system will meet the customer's actual needs, while verification is concerned with whether the system is well-engineered, error-free, and so on. Verification will help to determine whether the software is of high quality, but it will not ensure that the system is useful. The distinction between the two terms is largely to do with the role of specifications. Validation is the process of checking whether the specification captures the customer's needs, while verification is the process of checking that the software meets the specification. Verification includes all the activities associated with the producing high quality software: testing, inspection, design analysis, specification analysis, and so on. It is a relatively objective process, in that if the various products and documents are expressed precisely enough, no subjective judgements should be needed in order to verify software. http://www.easterbrook.ca/steve/2010/11/the-difference-between-verification-and-validation/
Ideal Time	Like Work Units, Ideal Time excludes non-programming time. When a team uses Ideal Time for estimating, they are referring explicitly to only the programmer time required to get a feature or task done, compared to other features or tasks. Again, during the first few iterations, estimate history accumulates, a real velocity emerges, and Ideal Time can be mapped to real, elapsed time. Many teams using Ideal Time have found that their ultimate effort exceeds initial programmer estimates by 1-2x, and that this stabilizes, within an acceptable range, over a few iterations. On a task by task basis the ratio will vary, but over an entire iteration, the ratios that teams develop have proven to remain pretty consistent. For a given team, a known historical ratio of Ideal Time to real time can be especially valuable in planning releases. A team may quickly look at the required functionality and provide a high level estimate of 200 ideal days. If the team's historical ratio of ideal to real is about 2.5, the team may feel fairly confident in submitting an estimate of 500 project days. In fixed-bid scenarios, this kind of estimate can be reliable. http://www.agilebok.org/index.php?title=Ideal_Time
Information Radiator	An information radiator is a large display of critical team information that is continuously updated and located in a spot where the team can see it constantly. The term "information radiator" was introduced extensively with a solid theoretical framework in Agile Software Development by Alistair Cockburn.

	Information radiators are typically used to display the state of work packages, the condition of tests or the progress of the team. Team members are usually free to update the information radiator. Some information radiators may have rules about how they are updated. Whiteboards, flip charts, poster boards or large electronic displays can all be used as the base media for an information radiator. For teams new to adopting agile work practices the best medium is usually a poster board on the wall with index cards and push pins. The index cards have a small amount of information on each of them and the push pins allow them to be moved around. Information radiators help amplify feedback, empower teams and focus a team on work results. Too many information radiators become confusing to understand and cumbersome to maintain. If an information radiator is not being updated it should be reconsidered and either changed or discarded. Robert McGeachy recommended a list of information radiators to be a part of every team room. Apart from the standard Scrum Artifacts, his list included, • Team structure with who's on the team • Client Organization structure • High Level and Mid Level plans • Client Phase exit criteria • Team performance survey results • Risks • Recognition awards • Ground Rules http://www.agileadvice.com/archives/2005/05/information_rad.html
Internal Rate Of Return (IRR)	Internal rates of return are commonly used to evaluate the desirability of investments or projects. The higher a project's internal rate of return, the more desirable it is to undertake the project. Assuming all projects require the same amount of up-front investment, the project with the highest IRR would be considered the best and undertaken first. A firm (or individual) should, in theory, undertake all projects or investments available with IRRs that exceed the cost of capital. Investment may be limited by availability of funds to the firm and/or by the firm's capacity or ability to manage numerous projects. In more specific terms, the IRR of an investment is the discount rate at which the net present value of costs (negative cash flows) of the investment equals the net present value of the benefits (positive cash flows) of the investment. http://en.wikipedia.org/wiki/Internal_rate_of_return
Iteration And Release Planning	Release planning is the process of transforming a product vision into a product backlog. The release plan is the visible and estimated product backlog itself, overlaid with the measured velocity of the delivery organization; it provides visual controls and a roadmap with predictable release points.

Agile Certification Study Guide – Glossary

	http://www.netobjectives.com/files/Lean-AgileReleasePlanning.pdf
I.N.V.E.S.T.	Attributes of an effective user story: <table><tr><td>Independent</td><td>The user story should be self-contained, in a way that there is no inherent dependency on another user story.</td></tr><tr><td>Negotiable</td><td>User stories, up until they are part of a Sprint, can always be changed and rewritten.</td></tr><tr><td>Valuable</td><td>A user story must deliver value to the end user.</td></tr><tr><td>Estimable</td><td>You must always be able to estimate the size of a user story.</td></tr><tr><td>Sized appropriately or Small</td><td>User stories should not be so big as to become impossible to plan/task/prioritize with a certain level of certainty.</td></tr><tr><td>Testable</td><td>The user story or its related description must provide the necessary information to make test development</td></tr></table> http://www.agileforall.com/2009/05/new-to-agile-invest-in-good-user-stories/
Kaizen (New 2015)	Kaizen is the practice of continuous improvement. Kaizen was originally introduced to the West by Masaaki Imai in his book Kaizen: The Key to Japan's Competitive Success in 1986. Today Kaizen is recognized worldwide as an important pillar of an organization's long-term competitive strategy. http://www.kaizen.com/about-us/definition-of-kaizen.html
Kanban Boards	A Kanban means a ticket describing a task to do. A Kanban Board shows the current status of all the tasks to be done within this iteration. The tasks are represented by cards (Post-It Notes), and the statuses are presented by areas on the board separated and labeled with the status. This Kanban Board helps the team understand how they are doing well as well as what to do next and makes the team self-directing.

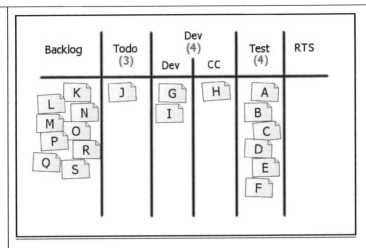

http://www.infoq.com/articles/agile-kanban-boards

http://blog.brodzinski.com/2009/11/kanban-story-kanban-board.html

Kano Model	The Kano model offers some insight into the product attributes which are perceived to be important to customers. The purpose of the tool is to support product specification and discussion through better development team understanding. Kano's model focuses on differentiating product features, as opposed to focusing initially on customer needs. Kano also produced a methodology for mapping consumer responses to questionnaires onto his model. The model involves two dimensions: • Achievement (the horizontal axis) which runs from *the supplier didn't do it at all* to *the supplier did it very well*. • Satisfaction (the vertical axis) that goes from *total dissatisfaction* with the product or service to *total satisfaction* with the product or service. Dr. Noriaki Kano isolated and identified three levels of customer expectations: that is, what it takes to positively impact customer satisfaction. The figure below portrays the three levels of need: expected, normal, and exciting. **Expected Needs** Fully satisfying the customer at this level simply gets a supplier into the market. The entry level expectations are the *must* level qualities, properties, or attributes. These expectations are also known as the *dissatisfiers* because by themselves they cannot fully satisfy a customer. However, failure to provide these basic expectations will cause dissatisfaction. Examples include attributes relative to safety, latest generation automotive components such as a self-starter, and the use of all new parts if a product is offered for sale as previously unused or new. The *musts* include customer assumptions, expected qualities, expected functions, and other *unspoken* expectations. **Normal Needs** These are the qualities, attributes, and characteristics that keep a supplier in the market. These next higher level expectations are known as the *wants* or the *satisfiers* because they are the ones that customers will specify as though from a list. They can either satisfy or dissatisfy the customer depending on their presence or absence. The *wants* include *voice of the customer* requirements and other *spoken* expectations (see table below).

	Exciting Needs The highest level of customer expectations, as described by Kano, is termed the *wow* level qualities, properties, or attributes. These expectations are also known as the *delighters* or *exciters* because they go well beyond anything the customer might imagine and ask for. Their absence does nothing to hurt a possible sale, but their presence improves the likelihood of purchase. *Wows* not only excite customers to make on-the-spot purchases but make them return for future purchases. These are *unspoken* ways of knocking the customer's socks off. Examples include heads-up display in a front windshield, forward- and rear-facing radars, and a 100,000 mile warranty. Over time, as demonstrated by the arrow going from top left to 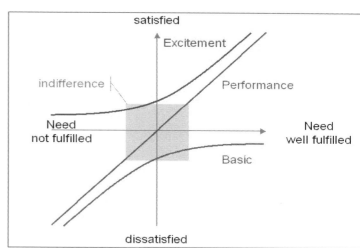 bottom right in the Kano model, *wows* become *wants* become *musts*, as in, for example, automobile self-starters and automatic transmissions. http://asq.org/learn-about-quality/qfd-quality-function-deployment/overview/kano-model.html http://en.wikipedia.org/wiki/Kano_model
KPIs in Agile (New 2015)	1. **Actual Stories Completed vs. Committed Stories** – the team's ability to understand and predict its capabilities. To measure, compare the number of stories committed to in sprint planning with the stories identified as completed in the sprint review. 2. **Technical Debt Management** – the known problems and issues delivered at the end of the sprint. It is usually measured by the number of bugs, but may also include deliverables such as training material, user documentation and delivery media. 3. **Team Velocity** – the consistency of the team's estimates from sprint to sprint. Calculate by comparing story points completed in the current sprint with points completed in the previous sprint; aim for +/- 10 percent. 4. **Quality Delivered to Customers** – Are we building the product the customer needs? Does every sprint provide value to customers and become a potentially releasable piece of the product? It's not necessarily a product ready to release but rather a work in progress, designed to solicit customer comments, opinions and suggestions. This can best be measured by surveying the customers and stakeholders. 5. **Team Enthusiasm** – a major component for a successful scrum team. If teammates aren't enthusiastic, no process or methodology will help. Measuring enthusiasm can be done by observing various sprint meetings or,

the most straightforward approach, simply asking team members "Do you feel happy?" and "How motivated do you feel?"
6. **Retrospective Process Improvement** – the scrum team's ability to revise its development process to make it more effective and enjoyable for the next sprint. This can be measured using the count of retrospective items identified, the retrospective items the team committed to addressing and the items resolved by the end of the sprint.
7. **Communication** – how well the team, product owner, scrum master, customers and stakeholders are conducting open and honest communications. Through observing and listening you will get indications and clues about how well everyone is communicating.
8. **Team's Adherence to Scrum Rules and Engineering Practices** – Although scrum doesn't prescribe engineering practices—unlike XP—most companies define several of their own for their projects. You want to ensure that the scrum team follows the rules your company defines. This can be measured by counting the infractions that occur during each sprint.
9. **Team's Understanding of Sprint Scope and Goal** – a subjective measure of how well the customer, product team and development team understand and focus on the sprint stories and goal. The goal is usually aligned with the intended customer value to be delivered and is defined in the acceptance criteria of the stories. This is best determined through day-to-day contact and interaction with the team and customer feedback.

http://pragmaticmarketing.com/resources/9-scrum-metrics-to-keep-your-team-on-track

Lead Time (New 2015)	Lead time is a term borrowed from the manufacturing method known as Lean or Toyota Production System, where it is defined as the time elapsed between a customer placing an order and receiving the product ordered. http://guide.agilealliance.org/guide/leadtime.html

Lean Development	Lean software development is a translation of lean manufacturing and lean IT principles and practices to the software development domain.
	
	1. **Eliminate Waste**
	   - Provide market and technical leadership - your company can be successful by producing innovative and technologically advanced products but you must understand what your customers value and you know what technology you're using can deliver
	   - Create nothing but value - you have to be careful with all the processes you follow i.e. be sure that all of them are required and they are focused on creating value
	   - Write less code - the more code you have the more tests you need thus it requires more work and if you're writing tests for features that are not needed you are simply wasting time
	2. **Create Knowledge**

- Create design-build teams - leader of the development team has to listen to his/her members and ask smart questions encouraging them to look for the answers and to get back with encountered problems or invented solutions as soon as possible
- Maintain a culture of constant improvement - create environment in which people will be constantly improving what they are working on - they should know that they are not and should not be perfect - they always have a field to improve and they should do it
- Teach problem-solving methods - development team should behave like small research institute, they should establish hypotheses and conduct many rapid experiments in order to verify them

3. **Build Quality In**
   - Synchronize - in order to achieve high quality in your software you should start worrying about it before you write single line of working code - don't wait with synchronization because it will hurt
   - Automate - automate testing, building, installations, anything that is routine, but do it smartly, do it in a way people can improve the process and change anything they want without worrying that after the change is done the software will stop working
   - Refactor - eliminate code duplication to ZERO - every time it shows up refactor the code, the tests, and the documentation to minimize the complexity

4. **Defer Commitment**
   - Schedule Irreversible Decisions at the Last Responsible Moment - you should know where you want to go but you don't know the road very well, you will be discovering it day after day - the most important thing is to keep the right direction
   - Break Dependencies - components should be coupled as loosely as possible to enable implementation in any order
   - Maintain Options - develop multiple solutions for all critical decisions and see which one works best

5. **Optimize the Whole**
   - Focus on the Entire Value Stream - focus on winning the whole race which is the software - don't optimize local inefficiencies, see the whole and optimize the whole organization
   - Deliver a Complete Product - teams need to have great leaders as well as great engineers, sales, marketing specialists, secretaries, etc. - they together can deliver great final products to their customers

6. **Deliver Fast**
   - Work in small batches - reduce projects size, shorten release cycles, stabilize work environment (listen to what your velocity tells you), repeat what's good and eradicate practices that creates obstacles
   - Limit work to capacity - limit tasks queue to minimum (one or two iterations ahead is enough), don't be afraid of removing items from the queue - reject any work until you have an empty slot in your queue
   - Focus on cycle time, not utilization - put in your queue small tasks that cannot clog the process for a long time - reduce cycle time and have

Agile Certification Study Guide – Glossary

	fewer things to process in your queue 7. **Respect People** • Train team leaders/supervisors - give team leaders the training, the guidance and some free space to implement lean thinking in their environment • Move responsibility and decision making to the lowest possible level - let your people think and decide on their own - they know better how to implement difficult algorithms and apply state-of-the-art software frameworks • Foster pride in workmanship - encourage passionate involvement of your team members to what and how they do (EXAM TIP: Lots of questions about Lean, XP and Scrum principles) http://agilesoftwaredevelopment.com/leanprinciples
Learning Cycles **(New 2015)**	The Build–Measure–Learn loop emphasizes speed as a critical ingredient to product development. A team or company's effectiveness is determined by its ability to ideate, quickly build a minimum viable product of that idea, measure its effectiveness in the market, and learn from that experiment. In other words, it's a learning cycle of turning ideas into products, measuring customers' reactions and behaviors against built products, and then deciding whether to persevere or pivot the idea; this process repeats as many times as necessary. The phases of the loop are: Ideas –> Build –> Product –> Measure –> Data –> Learn https://en.wikipedia.org/wiki/Lean_startup
Minimally Marketable Feature (MMF)	Start by identifying your product's most desirable features. Prioritize their value, either through a formal technique or by subjectively ranking each one against the others. Once you've done so, plan your releases around the features. Release the highest-value features first to maximize their return. To accelerate delivery, have your entire team collaborate on one feature at a time and perform releases as often as possible. http://jamesshore.com/Articles/Business/Software%20Profitability%20Newsletter/Phased%20Releases.html
Minimal Viable Product (MVP) **(New 2015)**	A Minimum Viable Product is that version of a new product which allows a team to collect the maximum amount of validated learning about customers with the least effort. http://leanstack.com/minimum-viable-product/
MoSCoW **(New 2015)**	The MoSCoW approach to prioritization originated from the DSDM methodology (Dynamic Software Development Method), which was possibly the first agile methodology (?) – even before we knew iterative development as 'agile'.

	MoSCoW is a fairly simple way to sort features (or user stories) into priority order – a way to help teams quickly understand the customer's view of what is essential for launch and what is not. MoSCoW stands for: • **M**ust have (or Minimum Usable Subset) • **S**hould have • **C**ould have • **W**on't have (but Would like in future) http://www.allaboutagile.com/prioritization-using-moscow/
Multistage Integration Builds	Multi-stage Continuous Integration (MSCI) is an extension of the common practice of shielding others from additional changes by only checking-in when individual changes have been tested and only updating an individual workspace when it's ready to absorb other people's changes. With MSCI, each team does a team-based continuous integration first and then cross-integrates the team's changes with the mainline on success. This limits project-wide churn and allows continuous integration to scale to large projects. http://www.accurev.com/multistage-continuous-integration.html
Net Present Value (NPV)	The difference between the present value of cash inflows and the present value of cash outflows. NPV is used in capital budgeting to analyze the profitability of an investment or project. NPV compares the value of a dollar today to the value of that same dollar in the future, taking inflation and returns into account. If the NPV of a prospective project is positive, it should be accepted. However, if NPV is negative, the project should probably be rejected because cash flows will also be negative. ***Formula*** Each cash inflow/outflow is discounted back to its present value (PV). Then they are summed. Therefore NPV is the sum of all terms, $$\frac{R_t}{(1+i)^t}$$ where *t* - the time of the cash flow *i* - the discount rate (the rate of return that could be earned on an investment in the financial markets with similar risk.); the opportunity cost of capital R_t - the net cash flow (the amount of cash, inflow minus outflow) at time *t*. For educational purposes, R_0 is commonly placed to the left of the sum to

	emphasize its role as (minus) the investment. http://en.wikipedia.org/wiki/Net_present_value
Negotiation	Negotiation, meaning "discussion intended to produce agreement", is fundamental to every software project. (And other projects too – my examples just happen to come from the software industry.) Developers and customers must reach agreement on what the system is supposed to do. A wise agreement will define achievable goals and meet the users' real needs. In Fisher and Ury's book, *Getting to Yes*, they call their approach "principled negotiation". It contains four key elements: <ul><li>Separate People from the Problem</li><li>Focus on Interests, not Positions</li><li>Invent Options for Mutual Gain</li><li>Use Objective Criteria</li></ul>Both parties would be better served to engage in dialog about their underlying interests. Such dialog is encouraged by agile processes. They promote discussion, provide better opportunities to explore interests, and avoid premature "lock in" of positions. Fisher and Ury's style of negotiation is not about winning and losing. It's about *everybody winning*. http://www.agilekiwi.com/peopleskills/the-power-of-negotiation/ Fisher, Roger and William Ury. Getting to Yes: Negotiating Agreement Without Giving In. New York, NY: Penguin Books, 1983.
Open Space Meetings	**Opening**: 1. Show the timeline (agenda), how the event breaks down into Opening, Marketplace of ideas, Break-out sessions, Closing. 2. Sponsor introduces the theme. Briefly. One or two minutes max. Long openings drain the energy of the meeting quickly. Get participants to work ASAP. 3. Facilitators introduce the principles and the format. Explain how the marketplace of ideas works. **Marketplace of ideas**: 1. Participants write 'issues' on pieces of paper. Preferably with bold markers, so they are easy to read from a distance. 2. Participants choose a timeslot for their topic on the agenda wall. 3. One by one, participants explain their issue to the others, with the aim of drawing the right people to their break-out-session. http://www.agileopen.net/on-open-space

Agile Certification Study Guide – Glossary

Osmotic Communications	Osmotic communication means that information flows into the background hearing of members of the team, so that they pick up relevant information as though by osmosis. This is normally accomplished by seating them in the same room. Then, when one person asks a question, others in the room can either tune in or tune out, contributing to the discussion or continuing with their work. When osmotic communication is in place, questions and answers flow naturally and with surprisingly little disturbance among the team. http://alistair.cockburn.us/Osmotic+communication
Pareto Principle	Pareto's law is more commonly known as the 80/20 rule. The theory is about the law of distribution and how many things have a similar distribution curve. This means that *typically* 80% of your results may actually come from only 20% of your efforts! Pareto's law can be seen in many situations – not literally 80/20 but certainly the principle that the majority of your results will often come from the minority of your efforts. http://www.allaboutagile.com/agile-principle-8-enough-is-enough/
Parking Lot Charts	Parking Lot Charts summarize the top-level project status. A parking-lot chart contains a large rectangular box for each **theme** (or grouping of user stories) in a release. Each box is annotated with the name of the theme, the number of stories in that theme, the number of story points or ideal days for those stories, and the percentage of the story points that are complete. http://www.change-vision.com/en/visualizingagileprojects.pdf
Payback Period	The length of time required to recover the cost of an investment. http://www.investopedia.com/terms/p/paybackperiod.asp
Personas	A persona, first introduced by Alan Cooper, defines an archetypical user of a system, an example of the kind of person who would interact with it. The idea is that if you want to design effective software, then it needs to be designed for a specific person. For the bank, potential personas for a customer could be named Frances Miller and Ross Williams. In other words, personas represent fictitious people which are based on your knowledge of real users (EXAM TIP: Specific exam question on why use "extreme" personas) http://www.agilemodeling.com/artifacts/personas.htm
Planning Poker	The idea behind Planning Poker is simple. Individual stories are presented for estimation. After a period of discussion, each participant chooses from his own deck the numbered card that represents his estimate of how much work is involved in the story under discussion. All estimates are kept private until each

Agile Certification Study Guide – Glossary

	participant has chosen a card. At that time, all estimates are revealed (the card is played) and discussion in differences between the estimates begins. The goal is to keep discussing until variance on the estimate only varies by 1. They highest final number is used for the product backlog. http://store.mountaingoatsoftware.com/pages/planning-poker-in-detail
Pre-mortem (new 2015)	Also called a Futurespective: The pre-mortem activity is great for preparing for an upcoming release or challenge. With a different perspective, the activity guides the participants to talk about all that could go wrong. Then the conversation switches to a mitigation and action plan. http://riskology.co/pre-mortem-technique/
Process Tailoring	You can decide to tailor up, or tailor down. This is up to each organization to decide based on its own business needs. You just need to state your approach in your guidelines and then follow it.. Always tailor up if you want to increase control on your project, simplify your planning/ tailoring process, and run projects with appropriate agility, efficiency, and discipline. http://www.enterpriseunifiedprocess.com/essays/softwareProcessImprovement.html
Product Roadmap	Product/Portfolio planning is a key activity for the Agile Product Manager, which usually consists of planning and management of existing product sets, and defining new products for the portfolio. Now, in order to define the portfolio, the product manager has to develop a **product roadmap** in collaboration with her stakeholders that consists of new upcoming products and existing product plan updates based on the their current status. The product roadmap thus enables identifying future release windows and drives planning for tactical development. http://www.agilejournal.com/articles/columns/column-articles/2650-product-road-mapping-using-agile-principles
Progressive Elaboration	Progressive elaboration is defined by The PMBOK® Guide as continuously improving and detailing a plan as more detailed and specific information and more accurate estimates become available as the project progresses, and thereby producing more accurate and complete plans that result from the successive iterations of the planning process. 1. Decide on a release timebox for the project. This may be one week, two weeks, one month. Whatever your team is comfortable with. 2. Look at the requirements on a high level and have the team decide approximately what you can release in each release cycle. Since this is a high level, approximate estimate, you don't need to be too detailed. It's just there to provide a rough idea about how the releases will develop.

Agile Certification Study Guide – Glossary

	3. At every iteration planning meeting, sit with the customer/product owner and decide what you are going to do in that iteration. At this stage, you can ask for more details, and the team can come up with a more accurate estimate based on the details that you now know. 4. At the end of the iteration, update the high level overview with any new information that you now have. 5. Repeat steps 3 and 4 for every iteration. http://www.agilebok.org/index.php?title=Progressive_Elaboration
Refactoring	Refactoring (noun): a change made to the internal structure of software to make it easier to understand and cheaper to modify without changing its observable behavior. http://martinfowler.com/bliki/DefinitionOfRefactoring.html
Relative Prioritization and Ranking	You use planning, ranking, and priority to specify which work the team should complete first. If you rank user stories, tasks, bugs, and issues, all team members gain an understanding of the relative importance of the work that they must accomplish. Ranking and priority fields are used to build several reports. You rank and prioritize work items when you review the backlog for a product or iteration. http://msdn.microsoft.com/en-us/library/dd983994.aspx http://www.processimpact.com/articles/prioritizing.html
Relative Sizing using Story Points	Your team collaboratively estimates each user story in story points. In his book "Agile Estimation and Planning," Mike Cohn defines story points this way: "Story points are a unit of measure for expressing the overall size of a user story, feature or other piece of work." Story points are relative values that do not translate directly into a specific number of hours. Instead, story points help a team quantify the general size of the user story. These relative estimates are less precise so that they require less effort to determine, and they hold up better over time. By estimating in story points, your team will provide the general size of the user stories now and develop the more detailed estimation of hours of work later, when team members are about to implement the user stories. http://www.agilebok.org/index.php?title=Relative_Prioritization_or_Ranking Mike Cohn. 2005. Agile Estimating and Planning. Prentice Hall PTR, Upper Saddle River, NJ, USA
Research Story	Sometimes programmers won't be able to estimate a story because they don't know enough about the technology required to implement the story. In this case, create a story to research that technology. An example of a research story is "Figure out how to estimate 'Send HTML' story". Programmers will often use a spike solution (see Spike Solutions) to research the technology, so these sorts of stories

are often called spike stories.

Programmers can usually estimate how long it will take to research a technology even if they don't know the technology in question. If they can't even estimate how long the research will take, timebox the story as you do with bug stories. I find that a day is plenty of time for most spike stories, and half a day is sufficient for most.

http://jamesshore.com/Agile-Book/stories.html

Retrospectives

The meeting performed at the end of an iteration or Sprint in which the project team identifies opportunities for improvement in the next iteration, or in their Agile Process in its entirety.

Approach

1. Set the Stage
   - Lay the groundwork for the session by reviewing the goal and agenda. Create an environment for participation by checking in and establishing working agreements.
2. Gather Data
   - Review objective and subjective information to create a shared picture. Bring in each person's perspective. When the group sees the iteration from many points of view, they'll have greater insight.
3. Generate Insights
   - Step back and look at the picture the team created. Use activities that help people think together to delve beneath the surface.
4. Decide What to Do
   - Prioritize the team's insights and choose a few improvements or experiments that will make a difference for the team.
5. Close the Retrospective
   - Summarize how the team will follow up on plans and commitments. Thank team members for their hard work. Conduct a little retrospective on the retrospective, so you can improve too.

http://agile2007.agilealliance.org/downloads/handouts/Larsen_448.pdf
Esther Derby and Diana Larsen. 2006. *Agile Retrospectives: Making Good Teams Great*. Pragmatic Bookshelf.

Return On Investment (ROI)

A performance measure used to evaluate the efficiency of an investment or to compare the efficiency of a number of different investments. To calculate ROI, the benefit (return) of an investment is divided by the cost of the investment; the result is expressed as a percentage or a ratio. If an investment does not have a positive ROI, or if there are other opportunities with a higher ROI, then the investment should be not be undertaken.

The return on investment formula:

	$$ROI = \frac{(\text{Gain from Investment - Cost of Investment})}{\text{Cost of Investment}}$$ (EXAM TIP: The only finance question that I had was on one how the Product Manager manages ROI) http://www.investopedia.com/terms/r/returnoninvestment.asp
Risk Areas	Tom DeMarco and Tim Lister identified five core risk areas common to all projects in their book, Waltzing with Bears: <ul><li>Intrinsic Schedule Flaw (estimates that are wrong and undoable from day one, often based on wishful thinking)</li><li>Specification Breakdown (failure to achieve stakeholder consensus on what to build)</li><li>Scope Creep (additional requirements that inflate the initially accepted set)</li><li>Personnel Loss</li><li>Productivity Variation (difference between assumed and actual performance)</li></ul> http://leadinganswers.typepad.com/leading_answers/2007/04/the_top_five_so.html Tom DeMarco and Timothy Lister. 2003. Waltzing with Bears: Managing Risk on Software Projects. Dorset House Publ. Co., Inc., New York, NY, USA
Risk Adjusted Backlog	Risk Adjusted Backlog focuses on where investment needs to be undertaken, based on risk. The normal risk assessment database process will provide a decreasing list of priorities from the risk calculation: Potential Consequence x Likelihood. It may be necessary to make decisions on which of these should be dealt with first within each of the risk bands. Ideally one would fund and rectify all high and significant risks within the current financial year. However, constraints on both funding and the time to prepare and complete work may cause this ideal to be delayed. http://www.agilebok.org/index.php?title=Risk_Adjusted_Backlog
Risk Based Spike	Spikes, another invention of XP are a special type of story used to drive out risk and uncertainty in a user story or other project facet. Spikes may be used for a number of reasons: <ol><li>Spikes may be used for basic research to familiarize the team with a new technology of domain</li><li>The story may be too big to be estimated appropriately and the team may use a spike to analyze the implied behavior so they can split the story into estimable pieces.</li><li>The story may contain significant technical risk and the team may have to</li></ol>

	do some research or prototyping to gain confidence in a technological approach that will allow them to commit the user story to some future timebox. 4. The story may contain significant functional risk, in that although the intent of the story may be understood, it is not clear how the system needs to interact with the user to achieve the benefit implied. http://www.agilebok.org/index.php?title=Risk_based_Spike
Risk Burn Down Graphs 	Risk Burndown graphs are very useful for seeing if the total project risk is increasing or decreasing over time. It allows stakeholders to see instantly if we are reducing project risk. 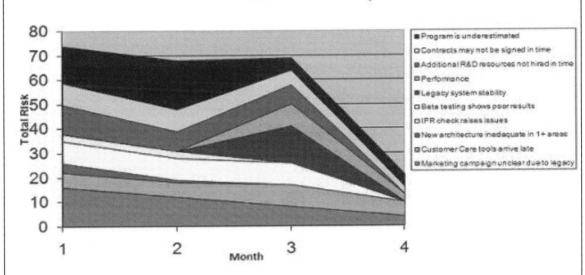 http://www.agilebok.org/index.php?title=Risk_Burndown_Graphs
Risk Exposure 	Probability of a risk multiplied by the impact (in days) if the risk occurs. (The damage done by a risk occurring.) http://blog.mountaingoatsoftware.com/managing-risk-on-agile-projects-with-the-risk-burndown-chart
Risk Maps **Risk Heat Maps**	Whether you are managing a large program or a small project you will need to constantly highlight the key risks to your project or program steering group. One of the best ways I know to visually display risks in a succinct manner is to use a Risk Map (also known as a Risk Heat Map). On the vertical axis you have the probably of a given risk occurring, that is, the likelihood that the risk will materialize and become an Issue. On the horizontal axis we have the impact that the risk will have on the project or program should it materialize. One of the benefits of this method of displaying risks is that it's easy to see how risky the program or project

Agile Certification Study Guide – Glossary

	is. If all the risks are clustered in the top right of the diagram then clearly your managing a very risky program or project 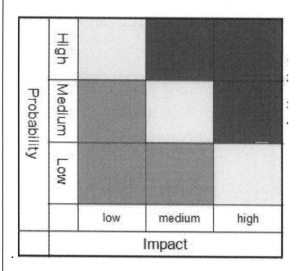 http://www.expertprogrammanagement.com/2009/06/visualise-risks-using-a-risk-map/
Risk Multipliers	Risk multipliers account for common risks, such as turnover, changing requirements, work disruption, and so forth. These risk multipliers allow you to set a date, estimate how many story points of work you'll get done, and be right. It's a simpler version of the risk curves you'll see in good books on estimating and project management. http://jamesshore.com/Blog/Use-Risk-Management-to-Make-Solid-Commitments.html
Scrum Ceremonies	There are four ceremonies - Sprint planning: the team meets with the product owner to choose a set of work to deliver during a sprint - Daily scrum: the team meets each day to share struggles and progress - Sprint reviews: the team demonstrates to the product owner what it has completed during the sprint - Sprint retrospectives: the team looks for ways to improve the product and the process http://www.scrumalliance.org/pages/scrum_101 (EXAM TIP: We did not go deeply into the Scrum Roles or Ceremonies in this Guide since this is the most commonly trained and well documented material. If you have not already attended a Scrum training, you should read Ken Schwaber's *Agile Project Management with Scrum*, or minimally, the latest Scrum

Agile Certification Study Guide –Glossary

	guide: http://www.scrumguides.org
Servant Leadership	There are several differences between Traditional projects and true Agile projects that—from a project management perspective—can best be summed up by the concept of self organization. In traditional projects, the project manager not only provides the vision of the team, but also directs and manages the team on the more detailed daily tasks by maintaining an up to date project plan. This usually results in a leadership style perhaps best described as "command and control." Agile projects on the other hand, still include the concepts of planning, managing the work, and providing status, but these activities are addressed collectively by the team, because at the end of the day they are the ones most familiar with what actually needs to happen to accomplish the project's goals. In this case, the Project Leader is in more of a support and facilitation role, similar in concept to Robert Greenleaf's idea of the Servant Leader. As Mike Cohn puts it in his Certified Scrum Master Class, the Project Leader's primary responsibilities are to "move boulders and carry water"—in other words, remove obstacles that prevent the team from providing business value, and make sure the team has the environment they need to succeed. One model often used to describe the leadership style needed on agile projects is the Collaborative Leadership model suggested by Pollyanna Pixton: • Make sure you have the **right people** on the project team. The right people are defined as those individuals who have passion about the goal of the project, have the ability to do the project, and are provided with the proper capacity, or time to work on the project. • **Trust First**, rather than waiting for people to prove their trustworthiness. • **Let the team members propose the approach** to make the project a success. After all, they are the ones who best know how to do the work. • **Stand back** and let the team members do their work without hovering over them continuously asking for status or trying to direct their activities, and provide support along the way to make sure nothing gets in the way of their success http://www.projectconnections.com/articles/092806-mcdonald.html Pollyanna Pixton, Niel Nickolaisen, Todd Little, and Kent McDonald. 2009. Stand Back and Deliver: Accelerating Business Agility (1st ed.). Addison-Wesley Professional
Shu Ha Ri (New 2015)	Shu-Ha-Ri is a way of thinking about how you learn a technique. The name comes from Japanese martial arts (particularly Aikido), and Alistair Cockburn introduced it as a way of thinking about learning techniques and methodologies for software development. The idea is that a person passes through three stages of gaining knowledge: • **Shu:** In this beginning stage the student follows the teachings of one master precisely. He concentrates on how to do the task, without worrying too much about the underlying theory. If there are multiple variations on

Agile Certification Study Guide – Glossary

	how to do the task, he concentrates on just the one way his master teaches him. - **Ha:** At this point the student begins to branch out. With the basic practices working he now starts to learn the underlying principles and theory behind the technique. He also starts learning from other masters and integrates that learning into his practice. - **Ri:** Now the student isn't learning from other people, but from his own practice. He creates his own approaches and adapts what he's learned to his own particular circumstances. http://martinfowler.com/bliki/ShuHaRi.html
Signal Card	English translation of the Japanese word, Kanban
Spike Solutions	A spike solution, or spike, is a technical investigation. It's a small experiment to research the answer to a problem. For example, a programmer might not know whether Java throws an exception on arithmetic overflow. A quick ten-minute spike will answer the question. http://jamesshore.com/Agile-Book/spike_solutions.html
Sprint Review	At the end of each sprint a sprint review meeting is held. During this meeting the Scrum team shows what they accomplished during the sprint. Typically this takes the form of a demo of the new features. The sprint review meeting is intentionally kept very informal, typically with rules forbidding the use of PowerPoint slides and allowing no more than two hours of preparation time for the meeting. A sprint review meeting should not become a distraction or significant detour for the team; rather, it should be a natural result of the sprint. Participants in the sprint review typically include the Product Owner, the Scrum team, the ScrumMaster, management, customers, and developers from other projects. During the sprint review the project is assessed against the sprint goal determined during the Sprint planning meeting. Ideally the team has completed each product backlog item brought into the sprint, but it is more important that they achieve the overall goal of the Sprint. http://www.mountaingoatsoftware.com/scrum/sprint-review-meeting
Story Maps	User story mapping offers an alternative for traditional agile planning approaches like the Scrum product backlog. Instead of a simple list, stories are laid out as a two dimensional map. The map provides both a high level overview of the system under development and of the value it adds to the users (the horizontal axis), and a way to organize detailed stories into releases according to importance, priority, etc. (the vertical axis). The map shows how every user story fits in the full scope. Releases are defined by creating horizontal slices of user stories, each slice is a

	release. For the first release, it is recommended to build a minimal set of user stories covering all user goals, so that you build a minimal but complete system to validate functionality and architecture early. http://blog.piecemealgrowth.net/working-with-user-story-mapping/
Sustainable Pace	To set your pace you need to take your iteration ends seriously. You want the most completed, tested, integrated, production ready software you can get each iteration. Incomplete or buggy software represents an unknown amount of future effort, so you can't measure it. If it looks like you will not be able to get everything finished by iteration end have an iteration planning meeting and re-scope the iteration to maximize your project velocity. Even if there is only one day left in the iteration it is better to get the entire team re-focused on a single completed task than many incomplete ones. Working overtime sucks the spirit and motivation out of your team. When your team becomes tired and demoralized they will get less work done, not more, no matter how many hours are worked. Becoming over worked today steals development progress from the future. You can't make realistic plans when your team does more work this month and less next month. Instead of pushing people to do more than humanly possible use a release planning meeting to change the project scope or timing. Fred Brooks made it clear that adding more people is also a bad idea when a project is already late. The contribution made by many new people is usually negative. Instead ramp up your development team slowly well in advance, as soon as you predict a release will be too late. A sustainable pace helps you plan your releases and iterations and keeps you from getting into a death march. Find your team's perfect velocity that will remain consistent for the entire project. Every team is different. Demanding this team increase velocity to match that team will actually lower their velocity long term. So whatever your team's velocity is just accept it, guard it, and use it to make realistic plans. http://www.extremeprogramming.org/rules/overtime.html
Tail Length	The tail is the time period from "code slush" (true code freezes are rare) or "feature freeze" to actual deployment. This is the time period when companies do some or all of the following: beta testing, regression testing, product integration, integration testing, documentation, defect fixing. http://www.allaboutagile.com/shortening-the-tail/
Task Boards	Similar to Kanban Boards, a task board tracks the progress of work that is part of an overall story.

	http://www.agilebok.org/index.php?title=Kanban_Task_Boards
Team Development Stages	**Tuckman's Group Development Model** **Forming** In the *first stages* of team building, the *forming* of the team takes place. The individual's behavior is driven by a desire to be accepted by the others, and avoid controversy or conflict. Serious issues and feelings are avoided, and people focus on being busy with routines, such as team organization, who does what, when to meet, etc. But individuals are also gathering information and impressions - about each other, and about the scope of the task and how to approach it. This is a comfortable stage to be in, but the avoidance of conflict and threat means that not much actually gets done. The team meets and learns about the opportunities and challenges, and then agrees on goals and begins to tackle the tasks. Team members tend to behave quite independently. They may be motivated but are usually relatively uninformed of the issues and objectives of the team. Team members are usually on their best behavior but very focused on themselves. Mature team members begin to model appropriate behavior even at this early phase. Sharing the knowledge of the concept of "Teams - Forming, Storming, Norming, Performing" is extremely helpful to the team. Supervisors of the team tend to need to be directive during this phase. The forming stage of any team is important because, in this stage, the members of the team get to know one another, exchange some personal information, and make new friends. This is also a good opportunity to see how each member of the team works as an individual and how they respond to pressure. **Storming** Every group will next enter the *storming* stage in which different ideas compete for consideration. The team addresses issues such as what problems they are really supposed to solve, how they will function independently and together and what leadership model they will accept. Team members open up to each other and confront each other's ideas and perspectives. In some cases *storming* can be resolved quickly. In others, the team never leaves this stage. The maturity of some team members usually determines whether the team will ever move out of this stage. Some team members will focus on minutiae to evade real issues. The *storming* stage is necessary to the growth of the team. It can be contentious, unpleasant and even painful to members of the team who are averse to conflict. Tolerance of each team member and their differences should be emphasized. Without tolerance and patience the team will fail. This phase can become destructive to the team and will lower motivation if allowed to get out of control. Some teams will never develop past this stage. Supervisors of the team during this phase may be more accessible, but tend to remain directive in their guidance of decision-making and professional behavior. The team members will therefore resolve their differences and members will be able to participate with one another more comfortably. The ideal is that they will not feel that they are being judged, and will therefore share their opinions and

	views. **Norming** The team manages to have one goal and come to a mutual plan for the team at this stage. Some may have to give up their own ideas and agree with others in order to make the team function. In this stage, all team members take the responsibility and have the ambition to work for the success of the team's goals. **Performing** It is possible for some teams to reach the *performing* stage. These high-performing teams are able to function as a unit as they find ways to get the job done smoothly and effectively without inappropriate conflict or the need for external supervision. By this time, they are motivated and knowledgeable. The team members are now competent, autonomous and able to handle the decision-making process without supervision. Dissent is expected and allowed as long as it is channeled through means acceptable to the team. Supervisors of the team during this phase are almost always participative. The team will make most of the necessary decisions. Even the most high-performing teams will revert to earlier stages in certain circumstances. Many long-standing teams go through these cycles many times as they react to changing circumstances. For example, a change in leadership may cause the team to revert to *storming* as the new people challenge the existing norms and dynamics of the team. http://en.wikipedia.org/wiki/Tuckman's_stages_of_group_development
Team Space	William Pietri put together a list of rules for great development spaces. Amongst the well documented suggestions like putting the team together, room for daily standup, enough whiteboards and information radiators other suggestions included, • Get collaboration-friendly desks – William suggested this as one of the big pitfalls. He mentioned that many companies would like to foster collaboration but end up having furniture which is hostile to it. • Minimize distractions – The recommended rules to minimize distractions for the development stations include no phones, no email or IM, no off-topic conversation, less foot traffic and executives stay on mute. • Only direct contributors sit in the room – No chickens and certainly not the receptionist nor the sales people who would mostly be on the phone. • Pleasant space – Good lighting, decent air, plants, decorations and snacks. http://www.infoq.com/news/2010/02/agile-team-spaces
Technical Debt Management	It's all "those *internal* things that you choose not to do now, but which will impede future development if left undone" [Ward Cunningham]. On the surface the application looks to be of high quality and in good condition, but these problems are hidden underneath. QA may even tell you that the application has quality and few defects, but there is still debt. If this debt isn't managed and reduced, the cost of writing/maintaining the code will eventually outweigh its

Agile Certification Study Guide – Glossary

	value to customers. In addition, it has a real financial cost: The time developers spend dealing with the technical debt and the resulting problems takes away from the time they can spend doing work that's valuable to the organization. The hard-to-read code that underlies technical debt also makes it more difficult to find bugs. Again, the time lost trying to understand the code is time lost from doing something more valuable. http://www.infoq.com/articles/technical-debt-levison **Managing technical debt** • **Starting captured debt.** Even if it is just by encouraging developers to note issues as they are writing code in the comments of that code, or putting in place more formal peer review processes where debt is captured it is important to document debt as it accumulates. • **Start measuring debt.** Once captured, placing a value / cost to the debt created enables objective discussions to be made. It also enables reporting to provide the organization with transparency of their growing debt. I believe that this approach would enable application and product end of life discussions to be made earlier and with more accuracy. • **Adopt standard architectures and open source models.** The more people that look at a piece of code the more likely debt will be reduced. The simple truth of many people using the same software makes it simpler and less prone to debt. http://theagileexecutive.com/2010/09/01/forrester-on-managing-technical-debt/
Test First Development 	Test-First programming involves producing automated unit tests for production code, before you write that production code. Instead of writing tests afterward (or, more typically, not ever writing those tests), you always begin with a unit test. For every small chunk of functionality in production code, you first build and run a small (ideally very small), focused test that specifies and validates what the code will do. This test might not even compile, at first, because not all of the classes and methods it requires may exist. Nevertheless, it functions as a kind of executable specification. You then get it to compile with minimal production code, so that you can run it and watch it fail. (Sometimes you expect it to fail, and it passes, which is useful information.) You then produce exactly as much code as will enable that test to pass. http://www.versionone.com/Agile101/Test-First_Programming.asp
Test-Driven Development	Test-Driven Development (TDD) is a special case of test-first programming that adds the element of continuous design. With TDD, the system design is not

	constrained by a paper design document. Instead you allow the process of writing tests and production code to steer the design as you go. Every few minutes, you refactor to simplify and clarify. If you can easily imagine a clearer, cleaner method, class, or entire object model, you refactor in that direction, protected the entire time by a solid suite of unit tests. The presumption behind TDD is that you cannot really tell what design will serve you best until you have your arms elbow-deep in the code. As you learn about what actually works and what does not, you are in the best possible position to apply those insights, while they are still fresh in your mind. And all of this activity is protected by your suites of automated unit tests.

You might begin with a fair amount of up front design, though it is more typical to start with fairly modest design; some white-board UML sketches often suffice in the Extreme Programming world. But how much design you start with matters less, with TDD, than how much you allow that design to diverge from its starting point as you go. You might not make sweeping architectural changes, but you might refactor the object model to a large extent, if that seems like the wisest thing to do. Some shops have more political latitude to implement true TDD than others.

http://www.versionone.com/Agile101/Test-First_Programming.asp |
| **Throughput (New 2015)** | Throughput is the amount of work items delivered in a given period of time (e.g. week, month, quarter).

http://berriprocess.com/en/todas-las-categorias/item/62-analiticas-lean-kanban-rendimiento |
| **Timeboxing** | Timeboxing is a planning technique common in planning projects (typically for software development), where the schedule is divided into a number of separate time periods (timeboxes, normally two to six weeks long), with each part having its own deliverables, deadline and budget. Timeboxing is a core aspect of rapid application development (RAD) software development processes such as dynamic systems development method (DSDM) and agile software development.

Scrum Timeboxes:
- Sprint Planning – 2 sessions, 1 hour per week of sprint
- Sprint Duration – 1-4 weeks
- Daily Scrums – 15 mins/day
- Sprint Review – 1 hours per week of sprint, 1 hour prep
- Sprint Retrospective – 3 hours per sprint

http://www.agilebok.org/index.php?title=Timeboxing |
| **Trade-Off** | Balancing the four constraints – compliance, cost, schedule, and scope – is not a |

Matrix	trivial task. However, just like the Agile Triangle, the Tradeoff Matrix used in Agile software development applies to IT. In its software development variant, the Tradeoff matrix is an effective tool to decide between conflicting constraints, as follows:

	Fixed	Flexible	Accept
Scope			X
Schedule	X		
Cost		X	

Rules:

- *Fixed* trumps *Flexible* trumps *Accepts*
- Each column can contain only one check mark
- Two check marks can't have the same priority

Note: The specific check marks in Table 1 are merely illustrative. Any three check marks that adhere to the rules above are legitimate. In fact, the three check marks represent the organization's policy decision as to what really matters.

http://theagileexecutive.com/tag/tradeoff-matrix/ |
| **Transition Indicator** | A transition indicator is a notification that a risk (i.e., something that will have a negative impact on the cost/schedule of the project *if* it occurs) has materialized and is in need of attention.

http://www.corvine.org/blog/archives/2004/08/waltzing_with_b.html |
| **Usability Testing (new 2015)** | Usability testing is a long-established, empirical and exploratory technique to answer questions such as "how would an end user respond to our software under realistic conditions?"

It consists of observing a representative end user interacting with the product, given a goal to reach but no specific instructions for using the product. (For instance, a goal for usability testing of a furniture retailer's Web site might be "You've just moved and need to do something about your two boxes of books; |

	use the site to find a solution.") http://guide.agilealliance.org/guide/usability.html
Use Cases	Use cases are sometimes used in heavyweight, control-oriented processes much like traditional requirements. The system is specified to a high level of completion via the use cases and then locked down with change control on the assumption that the use cases capture everything. Use cases attempt to bridge the problem of requirements not being tied to user interaction. A use case is written as a series of interactions between the user and the system, similar to a call and response where the focus is on how the user will use the system. In many ways, use cases are better than a traditional requirement because they emphasize user-oriented context. The value of the use case to the user can be divined, and tests based on the system response can be figured out based on the interactions. Use cases usually have two main components: Use case diagrams, which graphically describe actors and their use cases, and the text of the use case itself. http://en.wikipedia.org/wiki/Use_case
User Stories	A good way to think about a user story is that it is a reminder to have a conversation with your customer (in XP, project stakeholders are called customers), which is another way to say it's a reminder to do some just-in-time analysis. In short, user stories are very slim and high-level requirements artifacts. User stories are one of the primary development artifacts for Scrum and Extreme Programming (XP) project teams. A user story is a very high-level definition of a requirement, containing just enough information so that the developers can produce a reasonable estimate of the effort to implement it. http://www.agilebok.org/index.php?title=User_Stories_/_Backing
Value Stream Mapping	Value Stream Maps exist for two purposes: to help organizations identify and end wasteful activities. Finding problems and creating a more efficient process isn't easy; even the best organization can be made more efficient and effective. But bringing about substantive organizational change that actually eliminates waste is a tall order. It's comparatively easy to identify waste, but it's another matter entirely to stop waste from happening in the first place. Value Stream Maps can both sharpen an organization's skills in identifying waste and help drive needed change. But first things first: What a Value Stream Map is and how one can be intelligently produced. The examples and concepts that follow are based on applying a Value Stream Map to a software engineering organization, but these concepts are applicable to a wide range of settings. Value Stream Maps help us bring about organizational improvement, progress in our processes and methods, and most importantly, better software. Value Stream Maps can help both identify and stop waste in an organization

Agile Certification Study Guide – Glossary

	http://www.ibm.com/developerworks/rational/library/10/howandwhytocreateavaluestreammapsforswengineerprojects/index.html?ca=drs-
Velocity	Velocity is an extremely simple, powerful method for accurately measuring the rate at which teams consistently deliver business value. To calculate velocity, simply add up the estimates of the features (user stories, requirements, backlog items, etc.) successfully delivered in an iteration. There are some simple guidelines for estimating initial velocity prior to completing the first iteration but after that point teams should use proven, historical measures for planning features. Within a short time, velocity typically stabilizes and provides a tremendous basis for improving the accuracy and reliability of both near-term and longer-term project planning. Agile delivery cycles are very small so velocity emerges and can be validated very early in a project and then relied upon to improve project predictability. http://www.versionone.com/agile-101/agile-scrum-velocity/
Wide Band Delphi	Estimation method is a consensus-based technique for estimating effort 1. Coordinator presents each expert with a specification and an estimation form. 2. Coordinator calls a group meeting in which the experts discuss estimation issues with the coordinator and each other. 3. Experts fill out forms anonymously. 4. Coordinator prepares and distributes a summary of the estimates 5. Coordinator calls a group meeting, specifically focusing on having the experts discuss points where their estimates vary widely 6. Experts fill out forms, again anonymously, and steps 4 to 6 are iterated for as many rounds as appropriate. http://en.wikipedia.org/wiki/Wideband_delphi
WIP Limits	Limiting work in process (WIP) to match your team's development capacity helps ensure the traffic density does not increase the capacity of your team. The Kanban board will help you get to the right WIP limit as you become better at it. Without WIP limits you will continue to pile up partially completed work in the pipe thereby creating the phantom traffic jam. Adding to your WIP without completing anything just increases the duration of all tasks in the queue. If you are a product development shop, having a large duration (lead time) can significantly affect your company's profitability. http://www.kanbanway.com/importance-of-kanban-work-in-progress-wip-limits
Wireframes	A wireframe is a "low fidelity" prototype. This non-graphical artifact shows the skeleton of a screen, representing its structure and basic layout. It contains and localizes contents, features, navigation tools and interactions available to the

	user. The wireframe is usually: - black and white, - accompanied by some annotations to describe the behavior of the elements (default or expected states, error cases, values, content source...), their relationships and their importance, - often put in context within a storyboard (a sequence of screens in a key scenario) - refined again and again - used as a communication tool serving as an element of conversation and confirmation of "agile" user stories http://www.agile-ux.com/tag/wireframe/

Agile Roles

Scrum Roles	(EXAM TIP: I did not go deeply into the Scrum Roles or Ceremonies since this is the most commonly trained material. If you have not already attended a Scrum training, you should read Ken Schwaber's _Agile Project Management with Scrum_, or minimally, read the latest Scrum guide by him and Jeff Sutherland: http://www.scrumguides.org
Product Owner	The **product owner** decides what will be built and in which order. - Defines the features of the product or desired outcomes of the project - Chooses release date and content - Ensures profitability (ROI) - Prioritizes features/outcomes according to market value - Adjusts features/outcomes and priority as needed - Accepts or rejects work results - Facilitates scrum planning ceremony http://www.scrumalliance.org/pages/scrum_roles
Scrum Master	The ScrumMaster is a facilitative team leader who ensures that the team adheres to its chosen process and removes blocking issues. - Ensures that the team is fully functional and productive - Enables close cooperation across all roles and functions - Removes barriers - Shields the team from external interferences - Ensures that the process is followed, including issuing invitations to daily

Agile Certification Study Guide – Glossary

	scrums, sprint reviews, and sprint planning • Facilitates the daily scrums http://www.scrumalliance.org/pages/scrum_roles
Team	• Is cross-functional • Is right-sized (the ideal size is seven -- plus/minus two -- members) • Selects the sprint goal and specifies work results • Has the right to do everything within the boundaries of the project guidelines to reach the sprint goal • Organizes itself and its work • Demos work results to the product owner and any other interested parties. http://www.scrumalliance.org/pages/scrum_roles

Extreme Programming (XP) Roles

XP Coach	The XP Coach role helps a team stay on process and helps the team to learn. A coach brings an outside perspective to help a team see themselves more clearly. The coach will help balance the needs of delivering the project while improving the use of the practices. A coach or team of coaches supports the Customer Team, the Developer Team, and the Organization. The decisions that coaches make should always stem from the XP values (communication, simplicity, feedback, and courage) and usually move toward the XP practices. As such, familiarity with the values and practices is a prerequisite. The coach must command the respect required to lead the respective teams. The coach must possess people skills and be effective in influencing the actions of the teams. http://epf.eclipse.org/wikis/xp/xp/roles/xp_coach_60023190.html
XP Customer	The XP Customer role has the responsibility of defining what is the right product to build, determining the order in which features will be built and making sure the product actually works. The XP Customer writes system features in the form of user stories that have business value. Using the planning game, he chooses the order in which the stories will be done by the development team. He also defines acceptance tests that will be run against the system to prove that the system is reliable and does what is required. The customer prioritizes user stories, the team estimates them. http://epf.eclipse.org/wikis/xp/xp/roles/xp_customer_6D7CB91B.html
XP Programmer	The XP Programmer is responsible for implementing the code to support the user

	stories

http://epf.eclipse.org/wikis/xp/xp/roles/xp_programmer_D005E927.html- |
| **XP Programmer (Administrator)** | The XP Programmer (Administrator) role includes most of the traditional software development technical roles, such as designer, implementer, integrator, and administrator. In the administrator role, the programmer deals with establishing the physical working environment

http://epf.eclipse.org/wikis/xp/xp/roles/xp_system_administrator_92735060.html |
| **XP Tracker** | The three basic things the XP Tracker will track are the release plan (user stories), the iteration plan (tasks) and the acceptance tests. The tracker can also keep track of other metrics, which may help in solving problems the team is having. A good XP Tracker has the ability to collect the information without disturbing the process significantly.

http://epf.eclipse.org/wikis/xp/xp/roles/xp_tracker_AD8A6C9F.html |
| **XP Tester** | The primary responsibility of the XP Tester is to help the customer define and implement acceptance tests for user stories. The XP Tester is also responsible for running the tests frequently and posting the results for the whole team to see. As the number of tests grow, the XP Tester will likely need to create and maintain some kind of tool to make it easier to define them, run them, and gather the results quickly. Whereas knowledge of the applications target domain is provided by the customer, the XP Tester needs to support the customer by providing:

- Knowledge of typical software failure conditions and the test techniques that can be employed to uncover those errors.
- Knowledge of different techniques to implement and run tests, including understanding of and experience with test automation

http://epf.eclipse.org/wikis/xp/xp/roles/xp_tester_44877D41.html |

The Agile Manifesto

In February 2001, 17 software developers met at the Snowbird, Utah resort, to discuss lightweight development methods. They published the *Manifesto for Agile Software Development* to define the approach now known as agile software development. Some of the manifesto's authors formed the Agile Alliance, a nonprofit organization that promotes software development according to the manifesto's principles.

The Agile Manifesto reads, in its entirety, as follows:

We are uncovering better ways of developing software by doing it and helping others do it. Through this work we have come to value:

Agile Certification Study Guide – Glossary

> **Individuals and interactions** over processes and tools
> **Working software** over comprehensive documentation
> **Customer collaboration** over contract negotiation
> **Responding to change** over following a plan

That is, while there is value in the items on the right, we value the items on the left more.

The meanings of the manifesto items on the left within the agile software development context are described below:

- Individuals and Interactions – in agile development, self-organization and motivation are important, as are interactions like co-location and pair programming.
- Working software – working software will be more useful and welcome than just presenting documents to clients in meetings.
- Customer collaboration – requirements cannot be fully collected at the beginning of the software development cycle, therefore continuous customer or stakeholder involvement is very important.
- Responding to change – agile development is focused on quick responses to change and continuous development.

Twelve principles underlie the Agile Manifesto, including:

1. Our highest priority is to satisfy the customer through early and continuous delivery of valuable software.
2. Welcome changing requirements, even late in development. Agile processes harness change for the customer's competitive advantage.
3. Deliver working software frequently, from a couple of weeks to a couple of months, with a preference to the shorter timescale.
4. Business people and developers must work together daily throughout the project.
5. Build projects around motivated individuals. Give them the environment and support they need, and trust them to get the job done.
6. The most efficient and effective method of conveying information to and within a development team is face-to-face conversation.
7. Working software is the primary measure of progress.
8. Agile processes promote sustainable development. The sponsors, developers, and users should be able to maintain a constant pace indefinitely.
9. Continuous attention to technical excellence and good design enhances agility.
10. Simplicity--the art of maximizing the amount of work not done--is essential.
11. The best architectures, requirements, and designs emerge from self-organizing teams.
12. At regular intervals, the team reflects on how to become more effective, then tunes and adjusts its behavior accordingly.

http://agilemanifesto.org/

Looking for online practice exams for Agile certifications? Take our online practice exams for the PMI-ACP exam and for the Professional Scrum Master (PSM I) exam.

PMI-ACP CERTIFICATION SELF-STUDY TRAINING AND PRACTICE EXAM

This course is the perfect complement to our PMI-ACP Study Guide. All of the content in the Guide is represented in the Self-Study course and practice exam. The exams with the 25 questions as well as the one with 120 questions pull from a pool of over 370 questions so you will get a variety of questions each time to take it. We have reviewed hundreds of people's feedback on the exam and have reviewed all of the books referenced by PMI to help study for the exam. **To learn more about our self-study course**: http://bit.ly/ACPSelfStudy

Try out our sample questions: http://bit.ly/practice-pmi-acp

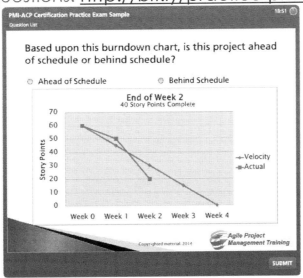

SCRUM MASTER CERTIFICATION PRACTICE EXAM

If you are looking to become a certified Scrum Master, Scrum.org offers the Professional Scrum Master certification, "PSM I". By taking our Scrum Master Certification Practice Exam you will be well prepared for the test. These questions will also help you to prepare for the Scrum portion of the PMI-ACP exam. To learn more about Scrum Master Certification and about this practice exam, click here: http://bit.ly/ScrumExam

References and Reccomended Reading

Agile & Iterative Development. Larman, Craig

Agile Estimating and Planning. Cohn, Mike

Agile Project Management with SCRUM. Schwaber, Ken

Agile Project Management: Creating Innovative Products – 2nd Edition. Highsmith, Jim

Agile Retrospectives: Making Good Teams Great. Derby, Esther; Larsen, Diana; Schwaber, Ken

Agile Software Development With Scrum, Schwaber, Ken; Beedle, Mike,

Agile Software Development: The Cooperative Game – 2nd Edition. Cockburn, Alistair

BecomingAgile: ...in an imperfect world. Smith, Greg; Sidky, Ahmed

Coaching Agile Teams. Adkins, Lyssa

Complexity and Management. D. Stacey, Ralph

Effective Project Management: Traditional, Agile, Extreme. Robert K. Wysocki

Exploring Scrum: The Fundamentals. Dan Rawsthorne with Doug Shimp

Extreme Programming Explained, Kent Beck, Addison

Kanban In Action. Marcus Hammarberg, Joakim Sunden

Kanban: Successful Evolutionary Change for your Technology Business. David J. Anderson

Lean-Agile Software Development: Achieving EnterpriseAgility. Shalloway, Alan; Beaver, Guy;Trott, James R.

The Software Project Manager's Bridge toAgility. Sliger, Michele; Broderick, Stacia

The Art ofAgile Development. Shore, James

User Stories Applied: ForAgile Software Development. Cohn, Mike

Software Engineering; A Practitioner's Approach. Pressman, Roger S.

Printed in Great Britain
by Amazon